IN THE SLIPSTREAM

IN THE SLIPSTREAM

A PILOT'S JOURNEY THROUGH WAR, PEACE, AND THE CREATION OF AMERICAN AIR POWER

CHRIS DELUSKY

ESSTHREE PUBLISHING LLC

IN THE SLIPSTREAM
Published by Essthree Publishing LLC
Detroit, MI USA

First Paperback Edition

ISBN: 979-8-9947851-2-6

Book Design by: Chris Delusky

All photographs by B.K. Watts (Author's collection) unless noted.

Author's Note

This story is built from records kept in motion. For nearly thirty years, Blanchard King Watts documented his life through flight logs, orders, and official forms, creating a near-continuous account of where and how he flew while larger historical events unfolded largely beyond his immediate view. History moved across maps and headlines. His experience was recorded from the air, one sortie at a time.

This book rests on a layered archive of voices and records. Foremost are Watts's flight logbooks, World War II war diaries, and official AF Form 5 entries. These are complemented by memoir fragments, photographs, and personal papers preserved in the author's collection. Together, these firsthand traces form the backbone chronology of missions, movements, and losses. Many items bear the marks of use and age: pencil entries clipped for space, pages darkened by field conditions, photographs with taped corners that have survived in albums for more than eighty years.

Those materials are supplemented by Colonel Watts's later memoir, *The Rise and Demise of the JOKER Squadron*, which captures his wartime experience through the lens of hindsight and memory. That manuscript is included here, as written, in the appendix.

Set against this personal record is the contemporary documentary archive: unit combat digests, official biographies, In Memoriam rolls, and official orders housed at the Air Force Historical Research Agency at Maxwell Air Force Base. Published operational histories provide the broader campaign frame. British official histories issued by His Majesty's Stationery Office, alongside parallel U.S. Army Air Forces and U.S. Army histories, add perspective from the wider Allied effort.

Where personal recollection and documentary records diverge, the book identifies the tension and cites both. For chronology and order of battle, the narrative privileges dated logs, orders, and official histories. Family memory and later oral history are treated as context. They supply tone, texture, and sometimes motive, but not the final word on time and place.

"**He was a man, take him for all in all,
I shall not look upon his like again.**"
~Hamlet 1.2.187–188

This book is dedicated to the memory of BK and Jeanne Watts, to the entire Watts and Delusky families, and to all branches of the service and their personnel who put everything on the line during both war and peace.

Blanchard King "BK" Watts, Col. USAF (Ret.), 1919–2010
Helen Jeanne Johnson Watts, 1923–2014

May their courage and sacrifice never be forgotten.

I am also very grateful to Gary W. Metz, author of *Last of the Randolph Blues*, whose encouragement at a critical moment helped ensure this story was finished.

Contents

Prologue..i

Part I: The Rise and Demise of JOKER Squadron
Chapter 1 Learning the Trade 1
Chapter 2 Crossing The Atlantic 9
Chapter 3 The Water at War 15
Chapter 4 Casablanca and the JOKERS 23

Part II: North Africa
Chapter 5 First Hard Lessons 35
Chapter 6 The Coil..43
Chapter 7 El Guettar: Striking Back.........................49
Chapter 8 Aftermath and Advance 57
Chapter 9 The Proving Ground 65
Chapter 10 The Fall of Tunis and Bizerte.....................73

Part III: Ciao Italia
Chapter 11 Across the Med 81
Chapter 12 Operation HUSKY....................................87
Chapter 13 The Race to Messina 93
Chapter 14 Celebrating in Sicily 99
Chapter 15 AVALANCHE..103
Chapter 16 From Naples to New Years......................109
Chapter 17 Villas, Volcanoes and Cassino Skies.........117
Chapter 18 The Beachhead and the Exit 123

Part IV: The CBI
Chapter 19 Hurry up and Wait 131
Chapter 20 Over the Hump......................................135
Chapter 21 120 More Days in China 145
Chapter 22 The Stage and the Smoke........................ 157

PART V: TEST

CHAPTER 23 CHANGE IS AFOOT 167

CHAPTER 24 INTO THE TEST ZONE...................171

CHAPTER 25 THE DESERT CRUCIBLE........................... 181

PART VI: THE COLD WAR

CHAPTER 26 BETWEEN WARS........................... 195

CHAPTER 27 COLD WAR PROFESSIONAL 201

CHAPTER 28 BACK TO CHINA 207

CHAPTER 29 STRATEGIC CURRENTS 215

CHAPTER 30 WAR COLLEGE 221

CHAPTER 31 THE LAST COCKPIT...........................227

CHAPTER 32 FOREIGN SERVICE...........................233

CHAPTER 33 DENOUEMENT 239

BIBLIOGRAPHY...........................245

APPENDIX A...........................253

APPENDIX B 271

INDEX 277

Prologue

The sea seemed gray and endless.

As dawn approached on the eve of the launches in November 1942, the carrier beneath their boots shuddered with every Atlantic swell. The world was already at war, and its reach stretched across continents.

HMS *Archer* was not a thoroughbred carrier but a converted Long Island-class merchant ship, her flight deck laid with creaking timber planks that complained under spray and weight. Oil, salt, and nerves hung heavily in the air. Operation TORCH was underway, the first Allied attempt to force a landing in Axis-occupied North Africa.

Four days earlier, word had spread through the fleet that another carrier, USS *Chenango*, had managed something few thought possible. She had launched dozens of P-40 Warhawks in a single day toward the Moroccan coast. Most of the planes made it. A handful had tumbled into the Atlantic, others cracked up on landing. In after-action signals, Western Task Force and Navy commanders called the effort a success. The improbable, it seemed, could be done.

None of that steadied the hearts here aboard *Archer*. She was smaller, rougher, and hastily converted. Her deck groaned under the added weight of the thirty-six replacement P-40s lashed to her deck, and her catapults had already balked at their burden. These were green pilots, most of whom had earned their wings only that summer, flying fighters never designed for carrier decks at all. Whatever *Chenango* had achieved belonged to another ship, and another morning. For the waiting pilots aboard *Archer*, that precedent offered little comfort.

Second Lieutenant Blanchard King "BK" Watts was only twenty-three years old. Five months earlier he had sweated under cadet discipline in the Texas sun. Now he stood on salt-stung planks beside British sailors who looked at him askance, one pilot among dozens about to hurl themselves into something neither they nor their airplanes had truly been meant to survive.

Before sailing, they had existed only on paper. In a wardroom off Norfolk, Virginia, their commanding officer, Major Philip Cochran, had chalked out a working roster. Names, pairings, who flew with whom. It was an exercise in order imposed on uncertainty, a list meant to move men efficiently rather than to define who they were.

Now the Atlantic was crowded with ships. Allied convoys filled the approaches to North Africa. Their columns were threaded by Royal Navy cruisers, destroyers, and escort carriers probing ahead for U-boats. George S. Patton's Western Task Force formed the American core of that movement, the only landing force to sail directly from the United States. The fleet was assembled in haste and secrecy, carrying men who had crossed an ocean before knowing where or when they would be committed.

For the men and pilots aboard *Archer,* the war was no longer abstract, it had narrowed to the distance between the deck and the sea below it.

<h1 style="text-align:center">PART I
THE RISE AND DEMISE OF JOKER SQUADRON</h1>

The Squadron did not begin as a unit.

They began in training, at a moment when the pipeline was still filling from all over the country, with the war already pulling men forward. Wings were earned, checklists completed, and aircraft assigned, even as combat experience lagged behind urgency.

Pulled from their current billets, they were gathered and reassigned. Not to a squadron with history or depth, but to a massive task that needed pilots and airplanes. Orders put them aboard a ship before any established military method existed for what they were going to be asked to do.

They flew from a carrier never meant to launch their aircraft, toward a field not yet secured, under conditions that allowed little rehearsal and no returns. What bonded them together in that moment wasn't a formal structure, it was shared risk.

Chapter 1

Learning the Trade

Blanchard King Watts came from rural western North Carolina, from a part of the state shaped more by fields than factories. By the time he finished high school, work was familiar to him. So were the machines that needed coaxing rather than confidence, and the low, practical flying of crop-dusting runs that treated an airplane used as a tool rather than an ambition. None of it announced a future in aviation, and none of it exempted him from what came next.

In 1941, he entered Army Air Forces pilot training as a cadet at Hicks Field, outside Fort Worth, Texas. He was twenty-two years old.[1]

Fig. 1-1 BK Watts Summer 1941 preparing for a training flight at Hicks Field, TX (Author's Collection)

Like the other men assigned there, Watts arrived with the minimum qualifications required by the expanding service. He had passed the physical examination, scored adequately on aptitude tests, and logged enough experience to be considered trainable. None of it guaranteed success. Hicks existed to decide who advanced and who did not.

The field itself was temporary by design. Runways were cut from the prairie soil. Tents snapped in the wind. Hangars went up faster than Army doctrine could follow. The threat of washout was constant. A bad pattern, a mishandled stall, or a lapse of judgment, and a cadet disappeared from the schedule.

The photograph above places BK there in that moment. His helmet is tilted back, his chinstrap loose, his jacket creased from use. He looks like the others around him, dressed for flight before flight had been earned. Their aircraft lie beyond the frame, Fairchilds and Stearmans waiting along the line, their presence implied rather than displayed. Nothing in the image distinguishes him from the others training there. Nothing promises an outcome.

Over Hicks Field, the war remained distant. Europe burned beyond the horizon, and in the Pacific the situation worsened by the week. Above the Texas prairie, the sky stayed a hard, kiln-fired blue. Cadets learned, often clumsily, to become airmen under conditions that allowed little margin for interpretation.

The field spoke in textures. Dust skittered across the runways. Canvas tent walls snapped against their guy lines. Engines throbbed across the sky like distant thunder. Prop wash lifted grit that settled into teeth, boots, and notebooks. Instructors strode the flight line with clipboards and hard eyes, their commands rising in sharp bursts before being shredded by engine noise.

Too flat in the pattern or too slow on a stall, and an instructor's pencil moved across the page. Those who held course crossed an unseen boundary. Behind them lay drills and dust. Ahead stretched a narrow road, held up by hours aloft, each tenth logged as a proof of passage.

Lieutenant Pointer ran Hicks Field with a stopwatch and a dry wit.[1] He flipped his pencil end over end while the boys talked, and they learned to land their planes before that pencil stopped moving.

Cadets were clustered by administrative convenience rather than preference. Watts found himself scheduled alongside Verne White, William "Bill" Wendt, and Charles "Chuck" Wood, their names adjacent in the roll book and therefore adjacent in the air. At Hicks, alphabetical proximity did the work of selection.

White flew smoothly and quietly, the sort who wrote down wind and altimeter settings even when no one asked. Wendt laughed big, carried gum for takeoff, and pushed the pace whenever he could. Wood earned "Woodchuck" over a card game and thereafter stacked everything he touched into careful order. Watts fit nowhere neatly. He talked fast, worked quickly, and treated checklists as challenges to be beaten rather than rituals to be respected.

Fig. 1-2 BK Watts Summer 1941 training flight at Hicks Field, TX (Author's Collection)

From there, his logbook took over. It spoke the language of training hours: 0.7. 1.3. 1.8. No sentiment. No drama. Just time accumulating under instruction. The numbers began to matter less as totals and became thresholds. Each entry marked another increment of control, another fraction of uncertainty pared away, another step forward granted rather than assumed.[1]

Fig. 1-3 BK Watts Summer 1941 pictured in his Randolf Blues at Randolph Field, TX (Author's Collection)

The West Point of the Air

By the fall of 1941, Watts had advanced to Randolph Field outside San Antonio, known formally and informally as the "West Point of the Air." Its white stucco blocks and central water tower projected authority as deliberately as its runways radiated outward like compass points. Randolph existed to standardize its pilots. Enthusiasm was burned off. Precision replaced it.[1] [2]

Every climb was measured against a stopwatch. Every slip of rudder became a notation in a file. More than half of those who entered would not advance. Fear was tolerated only long enough to be reduced to reflex. Training stripped the cadets down to give responses that could be evaluated, compared, and discarded if they failed to meet standard. By the time Watts was settled at Randolph, the syllabus no longer assumed a peacetime horizon.[2]

Pearl Harbor (7 December 1941)

The attack on Pearl Harbor altered pilot training almost immediately. Programs designed for measured progression were compressed, class sizes expanded, and the interval between qualification and assignment narrowed. Syllabi were adjusted to emphasize throughput as much as proficiency. Standards formally remained in place, but margins shrank. Experience was no longer expected to precede deployment. It would be acquired later, under operational conditions, by men already committed to the task.[3]

On the line, a crew chief named Ben Moore kept the aircraft moving. He worked without ceremony, eating when he could, coffee boiled down to tar. When cadets hovered with explanations about rough running or heavy controls, Moore listened just long enough to decide whether the answer mattered. Once, after overhearing White telling BK a long account of vibration on climb out, he wiped his hands on a rag and said, "Or you can just fly the airplane." Then he stepped clear of the prop arc.

Moore offered nothing beyond that. If an aircraft met specification, it flew. If it did not, it stayed put. Instruction came from procedures, not conversation. Randolph allowed no other interpretation.

Kelly

Kelly Field, also outside San Antonio, followed Randolph. The aircraft were faster and the formations closed tighter. Twenty cadets crowded into a single traffic pattern, each either to demonstrate control or to be removed from it. Instructors circled overhead in AT-6s, watching for deviation rather than effort.[1] [2]

Lieutenant Harry Green ran the check rides at Kelly.[1] He trusted the aircraft and the checklist. He did not negotiate. His evaluations were brief and final.

Training expanded on the ground as much as in the air. Gunnery ranges imposed discipline through repetition. Instructors covered blackboards with fuel calculations and range equations. Gallons became minutes. Minutes became margins. Watts learned to think of flying as arithmetic rather than motion, a sequence of tolerances that narrowed with every decision.

Before his required solo, the field grew quiet in his head. The pattern was prescribed. Wind from the south at eight. Full-stop landing only. Mechanics checked the aircraft and stepped back. No instruction followed and none was really expected.

He flew the pattern as assigned, executed the landing, and taxied clear. The flight was logged and his training continued.[1]

Graduation

The spring sun of 1942 shone hard off the San Antonio parade ground where Watts's silver wings were pinned.[1] His name was called in full by Army cadence. His family, far away in North Carolina, would read it shortened from a telegram: son, brother, uncle, and now a second lieutenant of the Army Air Forces.

On graduation day, the camera caught them shoulder to shoulder: Verne White, William "Bill" Wendt, Charles "Woodchuck" Wood, and Watts. Four names in order and four different grins. The photo belongs here because they went into the war together, even when orders sent them apart.

Orders did what orders do. Some of the boys went to England, some to the Pacific, and a few to schools and ranges nobody bragged about. They shook hands and made light of it, boys playing at bravado because that was the script. Verne promised to write. Bill thumped BK on the shoulder and said he would save him a stick of gum. Woodchuck said, "Box checked," and squared the corners of a kit bag that was not his. They did not say the quiet thing out loud, that this was likely the last time all four would stand in the same patch of sun.

Watts's orders pointed east to the Atlantic side, toward an

invasion still cloaked in code names and cover stories. Operation TORCH was coming, though nobody even whispered it yet. He left Texas with wings pressed to his chest, a dog-eared logbook written in decimals, and a mechanic's voice in his head. Ben Moore had said it plainly: fly the plane. Readiness was always provisional, it was a bridge you only knew you had crossed when you looked back from the far side.

Figure 1-4. Newly minted pilots at Kelly Field graduation, 1 July 1942. From left: B.K. Watts, Wood, White, and Wendt. These four would soon join the ranks of the U.S. Army Air Forces, with Watts launching into combat just 4 months later.(Author's Collection)

Notes

1. Watts, Blanchard King. Pilot Flight Logbook, 1941–1945. Author's collection.
2. Maurer, Maurer. Aviation Cadet Training in World War II. Maxwell Air Force Base: USAF Historical Studies, 1967.
3. U.S. Army Center of Military History. Pearl Harbor: 7 December 1941. Washington, D.C.: CMH Pub 72-7.

CHAPTER 2

CROSSING THE ATLANTIC

BK's official flight records reveal the pivot every Army Air Forces cadet eventually faced: the move from the predictable circuits of Texas trainers to the blunt realities of front-line fighter aircraft. After nine months of training, Watts had moved from his oil-stained Randolph blues to operational flying in the P-40. He was stationed on the East Coast, flying nearly every day from a home base near Philadelphia, accumulating hours that placed him firmly in the replacement pipeline rather than among recent graduates.[1]

Mitchel and Manhattan

During the latter part of the mid-Atlantic summer, Watts was assigned to the 324th at Philadelphia Municipal Field. He flew routine training hops along the eastern seaboard, working the corridor south to Langley Field and back. Ferry legs and local sorties filled his logbook with repeated shorthand: depot circuits, test climbs, short flights measured in tenths. It was workmanlike flying, designed to refine habits rather than test limits.[2]

On 22 October, however, those orders changed. Watts was directed to Middletown, Pennsylvania. Five days later, what began as a routine local training flight carried an added instruction: If the aircraft proved airworthy, the pilots were not to circle home.

Instead, they were to proceed east. Satisfied with his aircraft, he crossed New Jersey and the Hudson River, the familiar farmland gave way to the dense geometry of New York City. They landed at Floyd Bennett Field in Brooklyn, the joint Navy-Army air station pressed against the shoreline east of Manhattan.[3] [4]

```
SPECIAL ORDER )  R E S T R I C T E D   HEADQUARTERS I FIGHTER COMMAND
             :                           MITCHEL FIELD, NEW YORK
NUMBER    284 ) E X T R A C T               30 OCTOBER 1942

     1. The following named Officers of the 33rd Fitr Gp, WP by best avail-
able means of transportation from Floyd Bennett Field, N.Y. to Mitchel Field
N.Y., thence to secret port of Embarkation.
                       Major PHILIP G. COCHRAN 0-22464
2nd Lt. LYNN R BISHOP       0-661180    2nd Lt. LASSITER THOMPSON    0-651251
2nd Lt. HASKELL E ARTERBURN 0-661177    2nd Lt. BLANCHARD K WATTS    0-651255
2nd Lt. BERRY J DUNCAN      0-790640    2nd Lt. HARRY R HAINES       0-651310
2nd Lt. CLYDE P ADAMS       0-790605    2nd Lt. OSCAR V HEARING      0-728657
2nd Lt. MARVIN E CARPENTER  0-727578    2nd Lt. JOHN S WALKER        0-651409
2nd Lt. CHARLES L COTTRELL  0-790633    2nd Lt. JACK E HAWKE         0-728655
2nd Lt. KENNETH R SMITH     0-661386  ? 2nd Lt. ROBERT P KANTNER     0-728676
2nd Lt. CLAUDE E MCCREIGHT  0-728694    2nd Lt. LESTER W BROCK       0-791066
2nd Lt. CHARLES R KING      0-791093    2nd Lt. HERBERT C SAYLES     0-791149
2nd Lt. ROBERT J CONNOR     0-659666  x 2nd Lt. SHIRLEY E GARDNER JR 0-659490
2nd Lt. BERT J GOULAIT      0-349111    2nd Lt. ARCHIE C KELLER      0-727479
2nd Lt. PHIL R MCMILLS      0-727516    2nd Lt. RICHARD D MURDOCK    0-727524
2nd Lt. HARRY E FRENCH      0-727438    2nd Lt. WILLARD D CROSS      0-724390
2nd Lt. WILLIAM P MCBRIDE   0-727506    2nd Lt. TOM A THOMAS, JR     0-654691
2nd Lt. LEOPOLDO V RODRIGUEZ C-664661   2nd Lt. GEORGE W NIGHTINGALE 0-791027
1st Lt. GEORGE MATUCH, JR   0-433563    2nd Lt. IRVING S SILVERSTEIN 0-791152
2nd Lt. HARVEY J CIBEL      0-791071  x 2nd Lt. RICHARD R CUTEERP    0-412620
2nd Lt. EDWARD J TOBIN      0-791167
   TDN. FD 31 P 431-01,02,03,07,08 A 0425-23. FD 34 P 434-02,03 A 0425-23.In
lieu of subsistence, a flat per diem of $6.00 is authorized while traveling
on official business and while absent from permanent station in accordance
with existing law and regulations.

  *           *              *              *              *

              By order of Colonel QUESADA:

                              R. B. TRAVIS,
                              Captain, Air Corps,
                              Adjutant.

OFFICIAL:

R. B. TRAVIS
Captain, Air Corps,
Adjutant.

This is a true copy:  Harry R. Haines Jr.
                      Harry R. Haines Jr.,
                      2nd Lt., Air Res.
```

Fig 2-1 Movement Orders, 30 October 1942, directing BK and other pilots aka "JOKER" Squadron, to "Secret Port of Embarkation". Annotations by BK Watts. (Author's Collection)

Here, their routine broke. The aircraft they had just delivered were quickly taken over and swallowed into hangars, prepared for shipment. Instead of flying back to Philly, the pilots themselves were bivouacked at Bennett, close enough to smell the salt air off Jamaica Bay. For a day or two they lingered, their job complete but the purpose of their stopover was kept deliberately quiet. Then, just as abruptly, fresh orders arrived directly from Colonel Quesada. They were transferred to Mitchel Field on Long Island, turning the fighters over formally. From there the Warhawks vanished altogether, routed on to a "secret port" and loaded aboard freighters bound overseas.

For the pilots, it was a brief interlude that glowed in their memory. They had flown their first one-way mission and delivered real combat machines into the pipeline. With the aircraft gone, they held passes that released them into New York City. Liberty trains clattered into Manhattan, past Broadway's neon and midnight plates at Woolworth's. Jukebox horns fogged smoke-hazed bars. Until then, it was the only moment when training had brushed directly against the machinery of war, a fleeting taste of purpose before the long grind resumed.

Embarkation

On 1 November,the passes stopped coming and the pilots reported to their "secret" port of embarkation. In New York Harbor, *HMS Archer* waited, merchant-hulled and hastily converted, her decks were narrow and perpetually wet. Thirty-six pilots crossed her gangways in quick succession as the ship slid downriver almost before their bags were stowed. Once clear of the harbor, *Archer* joined convoy UGF-2, pushing into the North Atlantic's gray maw.[3][5]

They met their commanding officer only after the ship was underway. Major Philip G. Cochran took the wardroom without ceremony and began assigning duties as if marking territory: gunnery, ordnance, signals. When his gaze settled on Watts, he paused. "Maintenance," he said. The room went still. A fighter pilot assigned to wrenches and grease drew a low murmur. Cochran smiled and moved on.[4]

"Is this a joke?" someone asked.

The laughter that followed broke the tension, and with it came a name. Chalk scraped across the roster board: JOKER. In that moment, the thirty-six men ceased to be anonymous replacement pilots and became a unit by usage rather than by order. The name held because it fit, born of a remark and an answering quip in a rolling wardroom.

Major Philip G. Cochran (1910–1979)

Fig 2-3 Phillip G Cochran By United States Army Air Force http://www.explorepahistory.com/displayimage.php?imgId=4169, Public Domain accessed 15 September, 2025.

- WWII Service: As Major in the Army Air Forces, served as operations officer with the 33rd Fighter Group during the North Africa campaign where he commanded JOKER.

- Nickname / Persona: Known as "Flip" or the "one-man air force" for his aggressive style and flair.

- Famed Commands: Later co-created the 1st Air Commando Group in Burma with John Alison (1943–44), pioneering composite air operations that combined fighters, transports, gliders, and light bombers.

- Pop Culture: The basis for Milton Caniff's comic-strip hero "Flip Corkin" in Terry and the Pirates.

- Postwar: Rose to Colonel, USAF (Ret.), then entered private business; remembered for his fearless reputation and larger-than-life persona.[6]

Ship's Work, Men's Work

The P-40 weapon systems had been packed in Cosmoline, a wax preservative applied in the factories to protect steel against salt and rust. A kindness during storage, but it became a liability at sea. *Archer* carried no American solvents. The pilots became, by necessity, their own mechanics.[45]

They improvised. Mess towels became rags. Safety wire served as scrapers. Coffee tins held whatever could be scrounged that passed as degreaser. Breech blocks came apart and went back together until the wax softened and the mechanisms ran free. Checklists were scrawled in grease pencil and erased only when a weapon cycled cleanly.[4]

Fatigue blurred the days. Pilots slept upright in chairs, ready rooms thick with smoke and varnish. BBC forecasts crackled over the ship's speakers. V-Mail forms were filled, crossed out, rewritten. Watts's final version, compressed for microfilm, was brief: *I'm where I'm supposed to be.*[4]

FIG 2-2 HMS *Archer* loaded with 36 P40 Warhawks bound for North Africa Nov 1942 Lt. Cdr. Geoffrey B. Mason, RN (Rtd.), "HMS *Archer* (D78) – Chronology," Service Histories of Royal Navy Warships in World War II, Naval-History.Net. Accessed September 26, 2025.

The Catapult

The catapult tested poorly. Some mornings it launched cleanly. Other days it stalled mid-cycle, sailors hammering it back to life with ball-peen hammers and curses. Everyone watching understood the implication. These systems had been designed for lighter aircraft flown from ships built for the purpose. *Archer* was neither a true carrier nor a ship meant to endure that kind of strain. As the days passed, the routine hardened. Wake with drills. Clean Cosmoline from the guns. Study maps. Walk through the launch sequence on a narrow deck. The pilots learned the hand signals of British sailors whose names blurred but whose motions became absolute. When the wind shifted, the air carried a faint trace of land. Africa lay ahead.[3] [5]

The briefings were stripped down to essentials. Launch in pulses. Hold the heading and watch your time. Trust your altitude and find the airfield. Destroyers would recover those who failed. Everything else depended on arithmetic and machinery. The Warhawks waited lashed to *Archer's* deck, along with thirty-six men carrying a name born of jest toward a sky where calculation would matter more than confidence.[3] [5]

Notes

1. Watts, Blanchard King. Individual Flight Records, August–September 1942 (Philadelphia; P-40), October–November 1942 (Mitchel Field and embarkation). Author's collection.
2. U.S. Army Air Forces. Pilot Training in the AAF, 1939–1945. Army Air Forces Historical Office, 1946.
3. Watts, Blanchard King. Orders issued under Col. Elwood R. Quesada, October 1942, including directive to a "secret port of embarkation." Reproduced document image. Author's collection.
4. Watts, Blanchard King. The Rise and Demise of the JOKER Squadron. Unpublished memoir, ca. 1980s. Author's collection.
5. Freeman, Roger. Air Phase of the North African Invasion. AAF Historical Study no. 5. Washington: Army Air Forces, 1946.
6. Life Magazine, March 22, 1943; Caniff, *Terry and the Pirates*; *NY Times*, Jan 6, 1979.

Chapter 3

The Water at War

From overhead, the Atlantic resolved into a sprawl of waves filled with gray iron. Convoys stretched in staggered blocks, columns of ships plowing east, froth stitched white behind them. To a pilot circling above, it resembled a slow-moving city. To the men below, it was an army afloat. Operation TORCH, unprecedented in scale, was underway, and Patton's Western Task Force threaded its way east toward Morocco. It was the only arm of the North African invasion force launched directly from the United States.[1]

Why North Africa?

The question lingered even then. Why North Africa?

Pearl Harbor had pulled the United States into the war just eleven months earlier, but the nation was still finding its war footing. The divisions were still green, and the relationships between commanders and their units had not yet been tested under the sustained pressure of combat. A direct assault against Hitler's Fortress Europe, whether along the French coast or through the Low Countries, was judged suicidal in 1942.

Stalin demanded relief for the Eastern Front, where the Red Army bled at a scale the West could scarcely comprehend. The British pressed for action, but they could ill-afford another Somme and Washington needed proof that American forces could land, fight, and hold ground overseas.

North Africa became the compromise. The operation was placed under the overall command of General Dwight D. Eisenhower, whose task was to coordinate a coalition war effort as much as to oversee the landings themselves. North Africa offered a place to commit American troops alongside their allies, secure the Mediterranean, and test command, logistics, and doctrine under fire. It was meant as rehearsal and risk in equal measure, a proving ground where an untested army might learn whether it could function as a modern fighting force.[2]

On 8 November 1942, three Allied task forces struck French North Africa. The Central Task Force under Lloyd Fredendall aimed for Oran. Kenneth Anderson's Eastern Task Force steamed toward Algiers. George S. Patton's Western Task Force, the only element to sail directly from American shores, converged on Casablanca, the prize of French Morocco.[3]

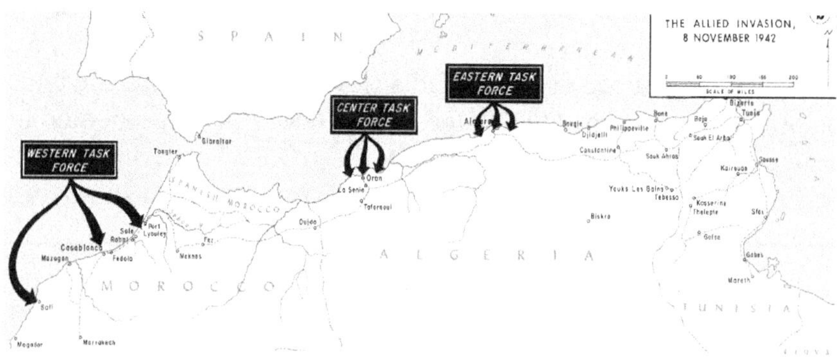

Figure 3-1 Allied landings in French North Africa, 8 November 1942, showing the three U.S. task forces of the Operation TORCH landing. Original U.S. Government map (public domain), via Wikimedia Commons accessed September 18, 2025

Patton commanded more than 35,000 soldiers and nearly 250 ships, a convoy that stretched across the horizon.[4] His landing sectors lay spread along the coast like cards laid on a poker table.

At Safi, U.S. Army Rangers stormed the port before dawn.

At Fedala, infantry splashed ashore under French fire until naval guns forced silence.

At Mehdia and Port Lyautey, P-40s launched inland to contest the airfield.[7]

Patton insisted on going ashore with the first waves, khakis pressed and revolver strapped high, wading into the surf in full view of his men.[8] Casablanca and its Vichy French defenders resisted until 11 November, when naval bombardment and infantry pressure compelled a ceasefire.[9]

Fig 3-2 Operation TORCH, November 1942. Major General George S. Patton, Jr., USA, prepares to go ashore in North Africa, shouting last-minute instructions. In foreground stands Rear Admiral John L. Hall, Jr., USN. Patton's Thompson submachine gun also visible. U.S. Navy photograph, released 21 December 1942. Courtesy of the Library of Congress.

HMS *Archer* in the Line

Stubby and slow among the larger ships of the convoy, HMS *Archer* rode in Patton's column.[10] Her cargo of thirty-six P-40s and the provisional JOKER Squadron, were, in truth, replacement pilots meant to join the fight once inland airfields were secured. Her American counterpart, USS *Chenango*, had already launched fighters toward Port Lyautey.[11]

The plan was audacious. Fighters never designed for carrier work would be launched toward contested fields with only minutes of fuel to spare.[13] *Chenango,* a true aircraft carrier, went first, shoving Warhawks toward Lyautey under fire. *Archer's* pilots waited. They heard the distant thunder of naval guns and felt the recoil shudder through the sea, but on *Archer's* deck, there was nothing to do but stand by.

Time to Launch

On 14 November, the order finally came.[13] *Archer* turned into the wind. Her deck was slick with spray, and the air was sharp enough that breath showed as men spoke. Everyone aboard understood the rule that governed what came next. Once airborne, there would be no ability to return to the deck. There were no arresting wires and no second attempts. They were flying directly into the war.

The aircraft were readied in sequence. Lashings came off one at a time. Deck crews moved with the practiced economy of men who had already seen what failure looked like. The P-40s sat forward, straight wings stretching nearly rail to rail, their weight leaving little tolerance for error.

Major Cochran did not wait. His Warhawk was walked into position and the checks were run without ceremony. Power came up, the catapult engaged, and the aircraft surged forward, lifting unevenly before steadying as it cleared the bow.

His voice came back almost immediately. "Move Watts up. Next."[13]

Watts had expected to lead the second wave, not to be pulled forward into the vacant slot. The aircraft now beneath him was not the one he had spent the passage learning and adjusting. That airplane would go with another pilot. This one was serviceable but unfamiliar, nothing wrong, just enough to demand attention. There was no discussion. The order stood.

The Flight

The loadout was stripped to essentials. Four of the six .50

caliber guns were removed, leaving fifty rounds apiece in the two that remained. Tool rolls and spares were left behind. Fuel was calculated at thirty minutes, if the sums proved true.[13] The entire navigation brief was reduced to a placard flashed by the deck crew. 080 degrees. Compass east for eighty miles.

Signals went up. Chocks came out. Gauges read green. The Allison V-12 settled into a steady roar. Over the radio, Cochran's final instruction came sharp and perfectly timed. "Don't stall!"

The catapult engaged. The aircraft resisted before yielding. The deck fell away. Spray hissed upward past the wings. For a moment, the Warhawk hung undecided between sea and air. BK, remembering Cochran's final instruction, let the stick out. Then, with the surf fast approaching, the prop bit into clean air and the horizon steadied. Wheels clattered up. *Archer* began to fall away behind him.

BK brought the aircraft around deliberately and circled the ship once. Below, sailors were already back at the catapult. Hammers rose and fell. Signal wands flashed in quick, frustrated patterns. The mechanism had failed again.

He circled a second time, long enough to confirm what the deck had already told him. There would be no immediate launches to follow. He was alone now.

He quickly checked his fuel. Thirty minutes. Enough to loiter briefly but not enough to wait. Cochran's aircraft was already far ahead, a dark shape pulling east

There would be no third circle.

Eighty Miles

Watts rolled wings level and turned east. Heading zero eight zero, and Eighty miles.[13] The arithmetic reduced everything else to irrelevance. 080 degrees for 80 miles, easy enough.

With the ship gone behind him, the world narrowed quickly. Sea filled the frame in every direction, broken only by shifting light and wind lines. The Warhawk settled into cruise. Engine temperature stabilized. Oil pressure stayed true. Each small confirmation mattered.

Time stretched. Five minutes passed, then ten. The horizon refused to change. Watts checked again. Compass steady. Altimeter holding. Fuel needle easing back. The numbers still worked.

At fifteen minutes, the light changed. The air grew warmer and the haze thickened ahead. The color of the water softened, losing its hard Atlantic steel. Then, a line appeared, faint at first and more suggestion than certainty, sharpening as he closed the distance.

The coast was real.

Mediouna

The sea fell away behind him. The engine settled into its rhythm, and the sky opened. Alone at altitude, the scale of things shifted. The convoy vanished into haze. The noise of the deck was gone. There was only the clock, the compass, and the steady pull of the aircraft.

He held the heading and watched the minutes tick down. There was time to think now, and just enough discipline not to. Into the unknown, he thought, and the phrase stayed with him, not as fear but as fact.

The air grew warmer as the coast approached. Haze thickened, then parted. Morocco came into focus with sharp lines and pale dust. Roads cut across fields. The outskirts of Casablanca slid beneath his wings.

He found Mediouna Airfield east of the city, scratched into the earth.[13] He chose to do a fly-by, low and fast, more curious than cautious, and was surprised to see more aircraft on the ground than he had expected. Relief came first, then confidence. He circled again, lined up, and set the Warhawk down.

When the Allison wound down, his logbook would show little more than a date and a time. What stayed with him were the details that did not fit in any column: the cardboard placard marked "080°," the frantic wands on *Archer's* deck, the joker card tucked under the seat, and the moment when the sea nearly claimed him.

The aircraft that carried him across was not his. The one he had tuned would fly under another pilot. This stranger's ship would fly too. He patted the cowling once, quietly. *Not my bird, but she got me here.*

Standing in Moroccan dust with salt still on his clothes, Blanchard King Watts understood he had crossed more than the Atlantic. Training lay behind him. Ahead lay combat.

Notes

1. Samuel Eliot Morison, History of United States Naval Operations in World War II, vol. 2, Operations in North African Waters, October 1942–June 1943 (Boston: Little, Brown, 1947), 45–52.
2. George F. Howe, Northwest Africa: Seizing the Initiative in the West (Washington, DC: Center of Military History, 1957), 3–10; C. P. Stacey, Official History of the Canadian Army in the Second World War, vol. II, The Canadians in Britain, 1939–1944 (Ottawa: Queen's Printer, 1955), 363–390.
3. Howe, Northwest Africa, 41–46.
4. Morison, Operations in North African Waters, 45–48; Howe, Northwest Africa, 42.
5. Howe, Northwest Africa, 42–43.
6. Morison, Operations in North African Waters, 48–52.
7. U.S. Army Air Forces, The Air Phase of the North African Invasion, November 1942–February 1943, AAF Historical Study No. 5 (Washington, DC: Army Air Forces, 1944), 29–31; Maurer Maurer, ed., Combat Squadrons of the Air Force, World War II (Washington, DC: USAF Historical Division, 1969), 165–67.
8. Martin Blumenson, Patton: The Man Behind the Legend, 1885–1945 (New York: William Morrow, 1985), 172–74.
9. Howe, Northwest Africa, 49–50; Morison, Operations in North African Waters, 60–63.
10. Morison, Operations in North African Waters, 60–63.
11. U.S. Army Air Forces, The Air Phase of the North African Invasion, 29–31; Morison, Operations in North African Waters, 62–63.
12. U.S. Army Air Forces, The Air Phase of the North African Invasion, 29–31.
13. Blanchard King Watts, The Rise and Demise of the JOKER Squadron, unpublished manuscript (ca. 1980s), 7–8. Author's collection.

CHAPTER 4

CASABLANCA AND THE JOKERS

Casablanca did not announce itself as a city. It came in fragments: the tang of salt and oil drifting inland from the surf, smoke hanging low where fires still smoldered among wreckage stacked along the quay. Five days earlier, the Western Task Force had taken the port, and the smell of battle still lingered, cordite and burned paint carried on the breeze. Offshore, the French battleship *Jean Bart*, incomplete and moored under Vichy command, lay scorched and holed where American naval gunfire and carrier strikes had silenced her single working turret. From the cockpit, the coastline looked half built and half broken.

With the P-40F settled onto the grass at Mediouna, BK realized it was his first landing in a combat zone. He had left *Archer* with less than thirty minutes of fuel. He sat for a beat with both hands still on the stick, breathing through the smell of hot oil and the coastal wind, then told himself to move.

As the propeller slowed, Major Cochran was already striding across the strip, cap tilted, boots splashed with dust.

"*Archer*. How'd she do?" He put a hand flat on the Warhawk's nose, as if feeling for a pulse. "Launches hold?"

"They beat the catapult back," BK said, loosening his harness. "Both of our kicks held then she broke down."

Cochran nodded once. That was enough.

They stood together and listened. Wind crossed the field.

Somewhere beyond the low hills, surf rolled in and out. Then an engine far off, a speck resolving into shape, another Warhawk was riding the coast inland.[1]

Figure 4-1 Lt. Robert Kantner's P-40F sinking after launch from HMS Archer, 13 Nov 1942 "Kantner" annotation by BK Watts. Note: Kantner's Autograph and inscription to BK. (Author's collection)

Survivors and the Sea

The strip filled in fits and starts. One aircraft. Then none. Then two together. By nightfall, thirty-three additional P-40s crowded Mediouna's dirt field. Three of their planes never made it.

Two catapult launches had broken sideways off *Archer's* deck and went into the Atlantic. The pilots, 2nd Lt. Robert Kantner and Lt. Oscar Hearing, scrambled alive from the cold water into the arms of a British destroyer. Kantner's Warhawk was photographed as it sank, tail lifting, wings spread wide and helpless before the sea folded it under.[2]

24

The third pilot was Lt. Marvin Carpenter. His launch had cleared cleanly. Then nothing. No return, no wreckage, only the Atlantic closing over. He was listed as presumed lost at sea.[3]

BK's memoir reduced the afternoon to a single line: "...We paced the strip all afternoon, sweating planes in. One, then two, then as many as seven or eight, but never an orderly appearing operation."[1] At dusk, thirty-three of thirty-six were on the ground, Watts and Cochran learning of the three missing pilots only after the men gathered in a makeshift operations room. Two recovered by destroyers. One gone entirely.

JOKER on the Move

Mediouna swelled overnight with tents, fuel drums, and ordnance until the field seemed to sag under the added weight. Cochran refused to let the informal squadron identity dissolve into the larger movement of the group. Within two days he led the thirty-three surviving pilots north to the Rabat-Cazes airfield.[4] [12]

At Cazes they inherited what the war left behind. French P-36 Mohawks were shoved into hangars. American tools seized under Vichy authority sat stacked in wire cages. Moroccan mechanics worked with quiet competence, willing to trade skill for rations. On the operations board, a grinning jester's face was sketched in chalk. A nickname hardened into habit. JOKER Squadron.[4] Cochran decided this would be the place to finish turning boys into fighter pilots.

Watts's logbook filled with one-hour entries. Behind the lines, nights meant soldering with flashlights clenched between teeth, coffee that smelled faintly of gasoline, and hands blistered raw on bomb fins and gun housings. What had begun as a joke was now sustained by work.[5]

Mud, Metal, and Generals

Cazes was mud disguised as a runway. On 20 November, General Carl Spaatz stood boots-deep in the Moroccan muck as a P-40 was buried nose-first off the Cazes airstrip. Minutes later, tanks clawed a bogged B-17 free so General Jimmy Doolittle could bully it back into the air. Improvised, absurd, and necessary, that was the rhythm of the days.[6] [7]

Fig 4-2 Lt. BK Watts at Rabat-Cazes Airfield. Provisional HQ for JOKER Squadron. Annotation by unknown. (Author's Collection courtesy of Jeanie Hemphill)

The Generals' presence signaled more than curiosity. Cochran's insistence that his hand-picked JOKER Squadron deserved to operate as a separate outfit had made ripples up the chain. To Spaatz and Doolittle, watching the 33rd FG try to fly out of Mediouna while half its replacements styled themselves a private squadron up the road in Rabat was a warning bell. Whether it was

Spaatz reminding the C.O. of the 33rd FG, Colonel William Momyer, to keep his officers aligned, or Doolittle himself probing the fissures, no one said aloud. The signal was clear: Headquarters had noticed.[6] [7] [9]

Figure 4-3 Col. William Momyer (left) and Maj. Philip Cochran, Rabat–Cazes Airfield, 7 Dec 1942. Momyer later signed the order that dissolved JOKER Squadron. Annotation by BK Watts. (Author's collection)

Collision in the Skies

Late November brought a harsher reminder of risk. Over Morocco, two JOKER Warhawks banked too close and collided in midair. 2nd Lt. Robert R. Smith never had a chance. His fighter fell in fire and dust, his name soon etched into the 33rd Fighter

Group's memorial rolls. His wingman, 2nd Lt. Herbert Sayles, survived the wreckage. By mid-December his name slipped quietly from the roster. He did not fly with the Jokers again.[8]

Empty cots multiplied. Familiar grins disappeared from the circle of tents. For a group that had only begun to believe in itself, the collision showed how quickly identity could vanish, one name erased, one call sign never answered again.

JOKER Squadron's Fate

On 7 December, Colonel Momyer arrived at Cazes. The photograph above preserves the moment. Cochran stands with his jaw set. Momyer is stiff, both men's boots thick with mud. The verdict was delivered without ceremony. JOKER Squadron would cease to exist. There would be no appeal. Pilots would be redistributed into existing fighter squadrons.[11]

JOKER Squadron had always existed within the 33rd Fighter Group. Its pilots were carried on the group's rolls even as the nickname took hold among the men. The order did not remove men from the 33rd. It rearranged them within it.

At higher levels, the logic was colder. By mid-December the push toward Tunisia had stalled against weather, logistics, and reorganized Axis resistance. Eisenhower needed results. Twelfth Air Force needed consistency. JOKER Squadron, however effective its cohesion, was still viewed as a replacement pool rather than a formation with a future.[9]

Cochran did not accept the decision quietly. Two days later he pulled Watts and another pilot and pointed their P-40s east along the coast toward Oran, where Twelfth Air Force headquarters had established itself. The flight was cold and silent. The purpose of the trip was plain: Cochran intended to save JOKER Squadron.[1]

Headquarters, however, was another world. Corridors ran straight. Floors shone. Staff officers carried neat folders instead of grease-stained rags. Cochran pressed his case. JOKER had launched from a carrier under hard odds. It had flown, fixed, and organized as a unit. The staff listened, nodded, and shuffled papers already filled with answers. Anomalies did not survive ledgers.[12]

Cochran did not return with them. Watts and the other pilot boarded a train west without their P-40s.[1] The third seat in the

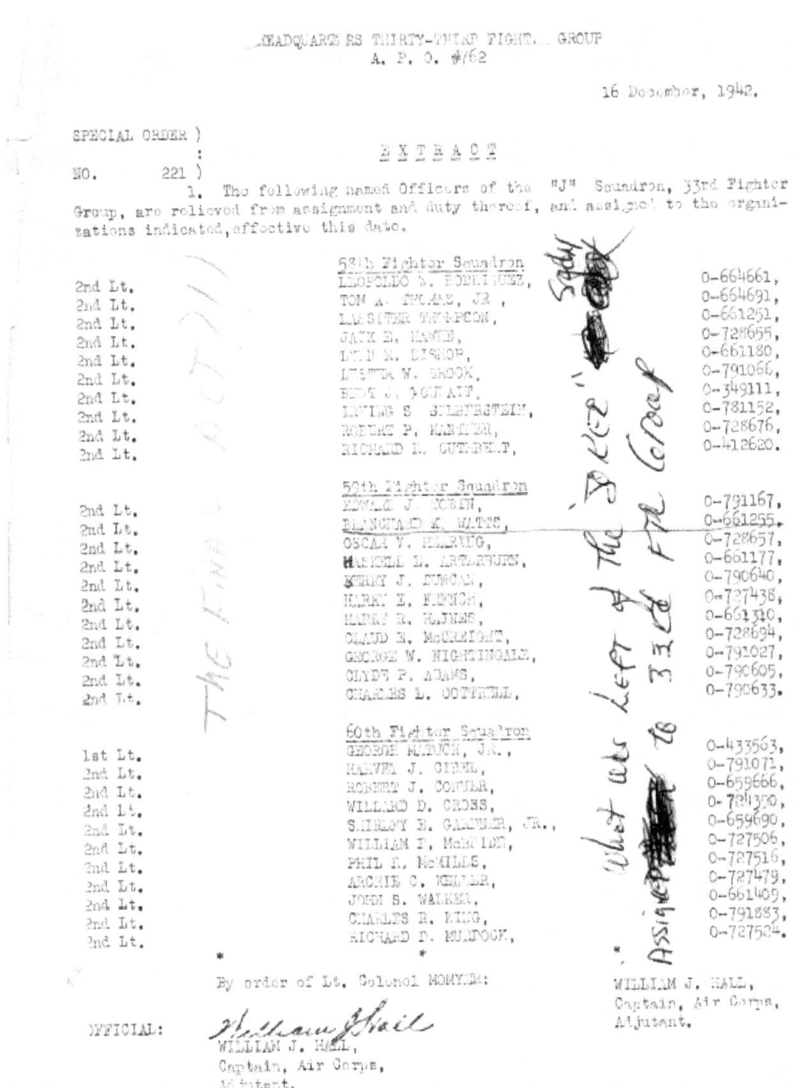

HEADQUARTERS THIRTY-THIRD FIGHTER GROUP
A. P. O. #62

16 December, 1942.

SPECIAL ORDER)
                :
NO.   221)   EXTRACT

          1. The following named Officers of the "J" Squadron, 33rd Fighter
Group, are relieved from assignment and duty thereof, and assigned to the organizations indicated, effective this date.

                              58th Fighter Squadron
2nd Lt.   LEOPOLDO B. RODRIGUEZ,   O-664661,
2nd Lt.   TOM A. THOMAS, JR ,   O-664691,
2nd Lt.   LASSITER THOMPSON,   O-661251,
2nd Lt.   JACK E. HAMER,   O-728655,
2nd Lt.   LYNN K. BISHOP,   O-661180,
2nd Lt.   LESTER W. BROOK,   O-791066,
2nd Lt.   BERT J. POTRAIT,   O-349111,
2nd Lt.   IRVING S. SILBERSTEIN,   O-781152,
2nd Lt.   ROBERT P. KAMINER,   O-728676,
2nd Lt.   RICHARD R. CUTHBERT,   O-412620.

                              59th Fighter Squadron
2nd Lt.   EDWARD J. ROBIN,   O-791167,
2nd Lt.   BLANCHARD K. WATTS,   O-661255,
2nd Lt.   OSCAR V. HERRING,   O-728657,
2nd Lt.   HASKELL B. ARTERBURN,   O-661177,
2nd Lt.   BOBBY J. DUNCAN,   O-790640,
2nd Lt.   HARRY E. FRENCH,   O-727438,
2nd Lt.   HARRY R. HAINES,   O-661340,
2nd Lt.   CLAUD E. McCRIGHT,   O-728694,
2nd Lt.   GEORGE W. NIGHTINGALE,   O-791027,
2nd Lt.   CLYDE P. ADAMS,   O-790605,
2nd Lt.   CHARLES L. COTTRELL,   O-790633.

                              60th Fighter Squadron
1st Lt.   GEORGE M. BUCH, JR.,   O-433563,
2nd Lt.   HARVEY J. ODELL,   O-791071,
2nd Lt.   ROBERT J. COWPER,   O-659666,
2nd Lt.   WILLARD D. CROSS,   O-784300,
2nd Lt.   SHELBY E. GARDNER, JR.,   O-659690,
2nd Lt.   WILLIAM F. McBRIDE,   O-727506,
2nd Lt.   PHIL R. McMILLS,   O-727516,
2nd Lt.   ARCHIE C. KELLAR,   O-727479,
2nd Lt.   JOHN S. WALKER,   O-661409,
2nd Lt.   CHARLES R. KING,   O-791883,
2nd Lt.   RICHARD D. MULDOCK,   O-727524.

          *   *   *

     By order of Lt. Colonel MOMYER: WILLIAM J. HALL,
                                                  Captain, Air Corps,
                                                  Adjutant.
OFFICIAL: William J Hall
                 WILLIAM J. HALL,
                 Captain, Air Corps,
                 Adjutant.

Fig 4-4 Special Order 221 from Colonel Momyer outlining assignments of the Jokers to their various new squadrons, ending JOKER squadron after less than 2 months. Annotations by BK Watts. (Author's collection)

compartment stayed empty. Cochran had been pulled into staff duty. His squadron dissolved by clerks rather than combat. Before long, the same headquarters that erased JOKER would send the Major east again, to Thelepte, to do in forward mud what he had already done once at Cazes.[13]

Orders and Absences

On 16 December 1942, Momyer signed the order. The language was sterile. "The following named officers of the 'J' Squadron…" Not JOKER, just the letter "J". There were thirty-two names. The squadron had launched with thirty-five pilots plus Cochran. Missing from the order were:

Lt. Marvin Carpenter, lost at sea
Lt. Robert R. Smith, killed in midair collision
Lt. Herbert Sayles, survived his collision but was never reassigned

The typewriter carried their absence as clearly as any marker.[11]

Notes

1. Watts, BK. The Rise and Demise of the JOKER Squadron, , unpublished manuscript (ca. 1980s), p. 7-8. Author's collection.
2. Ibid.
3. BK. Watts, Pilot Logbook, entry for 14 November 1942 ("Carpenter missing"). Author's collection; 33rd Fighter Group Combat Digest, In Memoriam, 1984 Reunion Booklet
4. Watts, B.K., Flight Logbook, Nov. 1942. Author's collection.
5. Twelfth Air Force. History of the Twelfth Air Force, 1942–1945. Washington, D.C.: HQ AAF, 1945, pp. 80–81.
6. Doolittle, James H., with Carroll V. Glines. I Could Never Be So Lucky Again. New York: Bantam Books, 1991, pp. 324–26.
7. 33rd Fighter Group Combat Digest. 1984 reunion booklet
8. Watts, B.K., The Rise and Demise of the JOKER Squadron, unpublished manuscript, p. 8. Author's collection
9. U.S. Army Air Forces. The Air Phase of the North African Invasion, November 1942–February 1943. Prepared by the Assistant Chief of Air Staff, Intelligence (Historical Division). Washington, D.C.: Headquarters, AAF, 1944, pp. 75, 79.
10. Howe, George F. Northwest Africa: Seizing the Initiative in the

West. United States Army in World War II. Washington, D.C.: Center of Military History, 1957, pp. 102–04.

11. Headquarters, 33rd Fighter Group. Extract Orders Relieving "J Squadron." Signed Col. William Momyer, 16 Dec 1942. Author's collection.

12. Robert T. Cooper, "The 33rd Fighter Group 'Fire from the Clouds'," Research Report 88-0600, Air Command and Staff College, Air University, Maxwell Air Force Base, AL, 16 February 1988, p. 23

13. Cooper, "The 33rd Fighter Group 'Fire from the Clouds'," p. 26.

Part II
North Africa

The beaches of Morocco had been only the beginning of the journey. Within weeks of Operation TORCH, the Allied foothold strained eastward toward Tunisia. This crawl began a campaign fought through winter mud, mountain passes, and the thin lifelines of airfields scraped from dirt and grass. What had seemed in November 1942 like a fast dash to eject the Axis instead became a grinding trial in logistics and endurance, as German veterans under Rommel and von Arnim stiffened their rapidly diminishing defense.

For the 33rd Fighter Group, and for the JOKER squadron alumni who had risked the one-way carrier launch from *HMS Archer*, North Africa was where war turned from experiment to ordeal. Here they learned to patrol, to escort, to strike convoys, and to bury their own. The forward bases at Thelepte and Youks-les-Bains were bomb-scarred dirt strips held on a shoestring, their P-40s patched with ration cans and kept flying through one of the worst winters Tunisia had recorded. On 15 January 1943, with the 33rd reduced from more than seventy operational aircraft to roughly thirty, repeated German and Italian attacks culminated in a climactic fight over Thelepte in which the 33rd's pilots destroyed eight Ju 88s without losing a single fighter. That became a day the War Department would later single out in awarding the Group a Distinguished Unit Citation.[1]

The losses at Maknassy and Sened in early 1943 were not simply lines in a log book but shocks that carved away the last illusions of combat as an adventure. North Africa was their battleground, shaping both the men and the squadrons for the campaigns still to come.

Chapter 5

First Hard Lessons

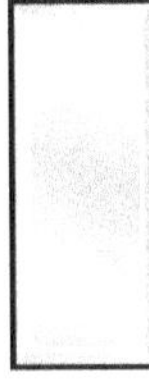

The morning Casablanca stopped sounding like a battlefield and started sounding like a port, everything shifted.[1]

Where once all eyes had looked west toward the beaches, headquarters's maps now stabbed east. Thick arrows ran from Morocco toward Tunisia, and convoys sketched in heavy pencil as the war reorganized itself in real time. The armistice with the Vichy French cracked open the gates of the city. Cranes groaned. Winches screamed. Trucks bucked across docks where French sentries had stood only weeks before. What had begun as a sprint into Morocco turned seemingly overnight into the slower, heavier grind of a new campaign.[2]

On the flight line outside Casablanca at Rabat, BK smelled the work before he saw it. Raw gasoline hung low in the damp air. Wet grain sacks bled dust into puddles. Grease and exhaust mixed with the urgency of machinery being pushed beyond any sensible margin. Still tasked with maintenance more often than flying, he checked his own airplane, fingers stiff as he worked straps, fittings, and control runs.[3] Behind him, freighters unloaded in a steady rhythm. Yesterday the war had arrived by carrier. Today it moved by convoy, grinding east through mud, train whistles cutting the mist at dawn.[4]

General George S. Patton moved through the docks in Casablanca with restless intensity, carrying a grease pencil rather than the ceremonial trappings later associated with him. He marked crates, pressed subordinates for impossible timelines, and treated delay as a moral failing rather than a logistical condition.[5] For Patton, momentum was not abstract theory but a discipline enforced minute by minute. That discipline depended on aircraft overhead. Fighters circling meant cranes could swing, dockworkers could work, and supply columns could move east without pause.[6]

Fig 5-1 NOV 1942 BK Watts at Rabat-Cazes Airfield with his P40. Annotation by BK Watts. (Author's Collection)

From Ceasefire to Cooperation

The ceasefire with the Vichy French felt brittle at first. Shoulders stayed tight at checkpoints. Salutes came stiff and late. Then, gradually, they loosened. Rifles leaned against sandbags. Rations were passed one way, cigarettes the other. French engines began to mix with American ones on adjoining flight lines. Mechanics who had been strangers days earlier started offering tools and fuel. Former Vichy supply depots became Allied supply caches almost overnight.[7]

Phil Cochran used to navigate this kind of transition easily.

He joked, gestured, and coaxed equipment out of warehouses still stamped with the French tricolor. To the pilots, it looked like sleight of hand. Sheds that seemed empty one day rolled out parts the next. Since Cochran had departed for another assignment, morale stopped orbiting personalities and settled back where it belonged, on the work itself. The 33rd remained busy because there was no alternative.[8]

When Colonel Momyer finally dissolved JOKER as a squadron on paper, folding its pilots and aircraft into the 58th, 59th, and 60th Fighter Squadrons, it felt less like a decision than an administrative inevitability. BK landed square in the middle, the 59th. It became his new home for as long as he was in the war. The ad-hoc unit that had begun as chalk marks on a carrier bulkhead vanished into tables of organization and strength returns. The *Archer* launch seemed to remain real only in their memory.[9]

Some units paid more for that shuffle than others. At Casablanca, the 59th learned what the ceasefire really cost. Held back while other elements moved east, the squadron was ordered first to train a Free French unit on P-40s and then to hand those same aircraft over to them. On 21 December, the entire base was marched to the ramp to watch a formal ceremony as French pilots taxied away in the Warhawks once assigned to the 59th. Officially, it was a gesture of Allied unity. On the line, it felt like they were being stripped.[10]

The men responded the only way they could. From invasion-damaged P-40s scattered around Casablanca, they scavenged engines, wings, and tail sections, and built five flyable fighters from wreckage. Those orphan planes never looked neat on paper, but they kept the squadron alive in the air. Pilots rotated through them for roughly an hour a day, just enough to remain current while awaiting orders east.[11]

First Missions East

BK's first flights beyond Morocco were limited and unspectacular by design. Escort work over Algeria, convoy cover stretched long and low across rain-darkened wadis and rail lines. These were not yet the sharp, violent sorties of Tunisia. They were apprenticeships. Radios sputtered and sulked. When they failed, geometry took over. Two wing dips to attack. Three dips to break north. Eyes outside the cockpit mattered more than any radio call that might be intercepted.[12]

Maps looked orderly on briefing tables. Rivers traced neat lines and roads appeared solid. From the air, however, everything was jagged. Pencil-thin rivers resolved into muddy trickles. Roads swallowed trucks axle-deep in mud. Bridges marked with confidence sometimes revealed themselves as shattered heaps. From the map table, the war looked manageable. From the cockpit, it felt improvised.[13]

On 4 January 1943, the 33rd FG moved east again, this time by transport to Thelepte, Tunisia.[14] Whatever the air planners intended, the shift pressed the Group closer to the front. For BK, it also clarified his status. He arrived as a replacement pilot, present but not yet flying combat.[15]

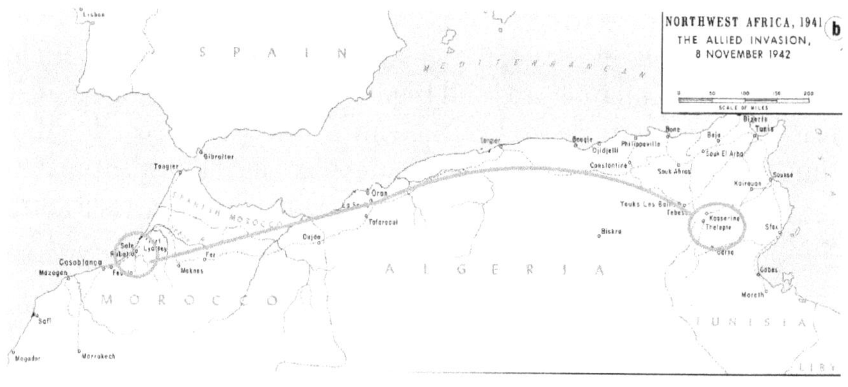

Fig. 5-2. 33rd Fighter Group movement from Rabat to Thelepte. Original U.S. Government map (public domain), Annotated by the author.

The forward strip at Thelepte was bulldozed clay and pierced steel planking (PSP) hammered into soaked earth. PSP glittered under rain like broken mirrors and clanged beneath engine run-ups, rattling tooth fillings loose before takeoff. When it rained, the field turned to glue. Tents flapped against iron stoves that coughed kerosene fumes and bitter coffee. Frost rimmed flight goggles each morning. There were no cockpit heaters. Pilots flew layered in silk, wool, leather, and fleece, with numb legs at altitude.[16]

Every dawn BK walked the strip, counting seams in the planking, testing puddles to see which had frozen overnight, memorizing where not to aim if he came back light on power. Only then did he climb into the cockpit, seal the canopy, and listen as the engine came up, the aircraft vibrating itself awake.

On the line, Sergeant Bernard "Bernie" Reed had already been working for an hour. He had crossed on *Archer* with JOKER and now owned BK's airplane as fully as any man could own a machine. Reed stripped his gloves to feel spark plugs, cursed metal that froze faster than fire could warm it, primed pumps until fuel ran clean, and safety-wired fittings by touch. When he finished, he rested a hand on the cowling, looked up at the cockpit, and said the same thing every morning.

"We will make it fly."[17]

It was not a speech. It was a statement of fact.

Fig 5-3 BK Watts's "Quarters" at Thelepte, Tunisia Jan 1943. Annotated by BK Watts. (Author's Collection)

Most days earned no glory. A patrol that came home without contact was success. An allied convoy that rolled unmolested was success. Radio logs that were a dull read were the best kind. Thelepte was a place where men learned patience before they learned violence.

By mid-January, the flight rhythm tightened. Aircraft turned faster. Briefings shortened. The logbook recorded times and headings, but didn't record the way fresh flak scars pocked fuselages or how ground crews repaired them without comment. Pilots dug their own shelters into stone and earth, pitching canvas tents low against wind and weather. At night they lay in those holes, listening to engines cough in the cold, waiting for the call that would finally move them from preparation into consequence.

Whatever was being built here was wound tight and growing tighter. Something would give. No one yet knew where.

Notes

1. George F. Howe, Northwest Africa: Seizing the Initiative in the West (Washington: Center of Military History, 1957), 88–90.
2. Samuel Eliot Morison, History of United States Naval Operations in World War II, Vol. II (Boston: Little, Brown, 1947), 75–77.
3. B. K. Watts, unpublished memoir, "The Rise and Demise of the JOKER Squadron," author's collection.
4. Morison, Naval Operations, 82.
5. Martin Blumenson, Patton: The Man Behind the Legend (New York: William Morrow, 1985), 174–76
6. Blumenson, Patton, 176; Howe, Northwest Africa, 91–92.
7. Howe, Northwest Africa, 93.
8. Watts, JOKER Squadron memoir.
9. Robert T. Cooper, "The 33rd Fighter Group: 'Fire From the Clouds,'" Air Command and Staff College, Maxwell AFB, 1988, chap. 5.
10. Cooper, "33rd Fighter Group," 27.
11. Cooper, "33rd Fighter Group," 28; James E. Reed, The

Fighting 33rd Nomads in World War II, Vol. I (Memphis: Historical Foundation, 1987).

12. Watts, pilot logbook, Jan 1943, author's collection.

13. AAF Tactical Reports, Tunisia, Jan 1943, NARA RG 18.

14. Watts, pilot logbook, 4 Jan 1943.

15. Watts, logbook and memoir, Jan 1943.

16. Reed, Fighting 33rd Nomads, Vol. I.

17. Watts, memoir; corroborated by Reed, Nomads.

CHAPTER 6

THE COIL

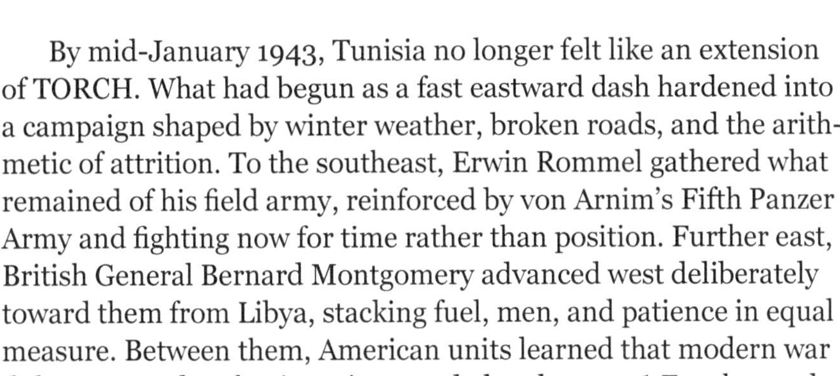

By mid-January 1943, Tunisia no longer felt like an extension of TORCH. What had begun as a fast eastward dash hardened into a campaign shaped by winter weather, broken roads, and the arithmetic of attrition. To the southeast, Erwin Rommel gathered what remained of his field army, reinforced by von Arnim's Fifth Panzer Army and fighting now for time rather than position. Further east, British General Bernard Montgomery advanced west deliberately toward them from Libya, stacking fuel, men, and patience in equal measure. Between them, American units learned that modern war did not reward enthusiasm it rewarded endurance.[1] For the 33rd FG, endurance was measured in available aircraft.

When the group moved east by transport to Thelepte on 4 January, the strip was already battered, and the PSP was hammered into the soaked clay. Fighters were needed overhead every hour the field remained usable, yet the airplanes arrived slowly and unevenly. Pilots arrived faster than the P-40s. Some pilots flew. Others waited, worked, and learned the war from the ground.[2]

The Flight Line

At Thelepte, the difference between assignment and belonging revealed itself on the ground, not in the air. Replacement pilots were present, necessary, and watched closely. They learned the strip by walking it, memorizing seams in the PSP seeking the low spots that flooded first, and the revetments that caught fragments when bombs fell short. They learned by standing beside men who had crossed the desert in combat, listening more than speaking.[3]

On 13 January, Watts noted simply that he had checked out in a P-39. The Airacobra was not his airplane and not yet his war, but the flight kept him current while P-40s trickled forward. The following days passed in proximity rather than action. Engines ran. Aircraft were shuffled. Missions launched with whoever was available. Combat happened overhead and sometimes on the airstrip itself.[4]

The pressure peaked on 15 January, when repeated Axis raids struck Thelepte and Youks-les-Bains. Patrols clawed upward through falling bombs and smoke. More than half of their aircraft burned on the ground while others lifted through smoke and dust. The Group's defense that day would later earn a Distinguished Unit Citation, but the paperwork could not capture the experience of standing under attack with too few fighters and no margin for error. Watts was there under the bombs, even if the logbooks would later remember other days more clearly.[5]

By 19 January, the waiting ended. Aircraft availability improved just enough, and Watts flew his first patrol in a P-40F. The next day, 20 January, he logged his first combat sortie over Maknassy, covering P-39s as they worked the roads and approaches. From that point forward, the pace accelerated without ceremony.[6]

The Maknassy-Mezzouna Corridor

By late January, the war had narrowed to a single stretch of road and valley, where traffic, flak, and fighters all converged. German 88mm cannons, carefully sited along wadis and junctions, reached up with increasing confidence. Radios failed often enough that formation geometry replaced speech. Two wing dips meant attack. Formation breaks were silent and became instinctive.[7]

Most sorties were short, an hour or a little more, but they stacked tightly. Armed reconnaissance flights blurred into escort. Escort slid into patrol. Pilots returned with the smell of cordite in their masks and dust ground into the seams of their flight jackets.

Thelepte barely held.[8]

On 28 January, Watts recorded another loss in his logbook without embellishment. Lt. Beard and Lt. White collided in mid-air. Both men bailed out over enemy-held ground and were later confirmed prisoners of war. The entry occupied a single line in his logbook. Its weight did not. Collision was a reminder that danger came from more than flak and fighters. The margin for error had narrowed everywhere.[9]

Losses

By 31 January, the pressure translated into more absence. Over the Maknassy-Mezzouna corridor, Messerschmitts dropped cleanly out of altitude and struck American flights moving along the road network. When the formations reassembled, two pilots from the 60th Fighter Squadron were missing: 2nd Lt. Harvey Cibel and 2nd Lt. Shirley Gardner Jr. Both were former JOKER pilots. Both had crossed into Africa the hard way, and their loss reached backward into the group's shared beginning. The war felt as intimate as ever.[10]

Figure 6-2 Damage to BK Watts' Warhawk wing fabric ripped by 88mm flak, 31 January 1943. Annotation by BK Watts.(Author's collection)

The following day, 1 February, brought no relief. Over Maknassy, another aircraft failed to return. Lt. Campbell was lost. Watts recorded it plainly in his log. There was no added commentary, no explanation. The accumulation itself was becoming the message.[11]

On 2 February, the bad news hit them hard.

That morning, BK flew escort for A-20 bombers over Faid. The mission itself appeared routine in his logbook. Elsewhere, it was anything but. The 59th Fighter Squadron was hit badly. Capt. Boone, Johnson, Smith, and Harry French did not return. French was a JOKER, one of Watts's own squadron mates, and his absence landed close.[12]

Two others lived by inches. Arterburn and Hilmo crash-landed. They were alive, but they were out of the sky, and that counted in its own way. Only Hutter came back to the field, shot up, the aircraft and the pilot carrying the same message.[12]

After 2 February, the salutes before takeoff were no longer casual gestures. They became deliberate and purposeful. When Watts flew, he raised one. When he remained on the ground, he returned it. No one needed to explained the change.[13]

The Coil Tightens

By early February, the front compressed. Axis armor had gathered beyond the passes. American units were shifted and reshuffled, and still learning how to hold ground under pressure. Far to the rear, Patton simmered in Casablanca, soon to be pushed forward.[14] At Thelepte, the men on the line did not need speeches to tell them what was coming. They could see it in the traffic below and in the way flak climbed faster each day.

BK's log for the period stayed spare: patrols, escorts, armed reconnaissance. Times and places, nothing more. What it did not record were the looks exchanged on the line, the pause before ground crews stepped back from spinning props, or the silence that followed roll call when names went unanswered.

The coil was set. It would snap soon enough.

46

Notes

1. George F. Howe, Northwest Africa: Seizing the Initiative in the West (Washington, DC: Center of Military History, 1957), 103–109; Williamson Murray and Allan R. Millett, A War to Be Won (Cambridge, MA: Belknap Press, 2000), 305–307.
2. Howe, Northwest Africa, 110–112.
3. Robert T. Cooper, "The 33rd Fighter Group: Fire From the Clouds," Air Command and Staff College monograph, chap. 5.
4. Blanchard K. Watts, Pilot Flight Logbook, January 1943 (author's collection).
5. War Department, General Orders No. 25 (7 April 1945), Distinguished Unit Citation, 33rd Fighter Group; Howe, Northwest Africa, 113–115.
6. Watts, Flight Logbook, 19–20 January 1943.
7. Twelfth Air Force Tactical Report No. 23, "Employment of Fighter Aircraft in Tunisia," 1944.
8. Howe, Northwest Africa, 116–118.
9. Watts, Flight Logbook, 28 January 1943.
10. Watts, Flight Logbook, 31 January 1943; 33rd Fighter Group Combat Digest, "In Memoriam," 1984 Reunion Booklet.
11. Watts, Flight Logbook, 1 February 1943.
12. Watts, Flight Logbook, 2 February 1943; 59th Fighter Squadron casualty summaries, February 1943.
13. Blanchard K. Watts, unpublished memoir notes (author's collection).
14. Martin Blumenson, Patton: The Man Behind the Legend (New York: William Morrow, 1985), 180–183.

CHAPTER 7

EL GUETTAR: STRIKING BACK

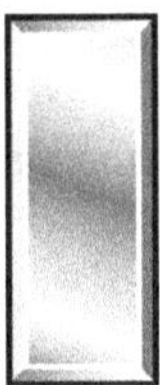

In early February 1943, newly promoted First Lieutenant Blanchard King Watts tightened the collar of his flight jacket against the desert morning chill as the 59th Fighter Squadron broke camp. For the first time since TORCH, the unit was ordered west, away from the narrowing fight in Tunisia and back toward Morocco.[2]

Fig 7-1 Feb 1943 Boys from the 59th in Thelepte preparing to move back to the west. Annotations by BK Watts. (Author's Collection)

The order came without ceremony. Trucks were allocated. Maps were folded and refolded. Engines were warmed long enough to be trusted, then shut down again. After weeks of pressure building eastward, the direction reversed. The 59th was moving west. No one called it a withdrawal. It was simply movement, and movement was enough. They learned soon enough that they were getting some R&R after the tremendous flurry of fighting over the previous 45 days.

The tents at Thelepte came down quickly, the same way they had gone up, with the practiced economy learned under threat. Propellers were wired down. Tool kits were lashed. Bedrolls were stuffed without fuss. A few hours later the column rolled out, leaving behind steel matting, spent fuel drums, and the marks of aircraft that had not left under their own power.[1]

The Safari

The road toward Agadir cut across scrub and low hills, far away from the Maknassy-Mezzouna corridor where losses had accumulated and the flying had begun to feel compressed. Dust trailed the trucks in a low haze. Conversation came and went. No one asked where they were headed next. The men had learned that questions rarely improved outcomes. They camped near the coast, where the air carried salt instead of cordite and the wind moved differently, lifting grit away rather than driving it deeper into skin and cloth. The change registered physically before it registered anywhere else. Shoulders settled. Hands stopped shaking when they were idle. Sleep came more easily, without calculation.

They went inland early, guided by local men who moved through the low hills with an ease born of familiarity. Rifles were carried without comment. Dogs threaded forward and back, alert but disciplined. For the first time in weeks, the pilots smiled without checking themselves. The smiles were small and unguarded, but they spread anyway. Jackets were unzipped against the morning chill. Someone said something that would not have landed well on the flight line. It landed here.

They stood close together before moving out, the way men do when spacing is instinctive and trust does not need to be demonstrated. What followed was quick and unremarkable in the telling. The animal broke cover. Shots were fired. The boar went down. No one ever fixed who fired the shot, and no one cared to. The guides took over as they always did, tying the hooves and bearing the weight forward with practiced ease.

Fig 7-2 & 7-3 Agadir Boar hunt 1943 (Top) the boys prepare with the guides. Bottom: The guides show off the fruits of their labors. Annotations by BK Watts. (Author's Collection)

The pilots walked alongside, hands empty now, talking again. Laughter came easier once the work was done. The weight they carried was real and indisputable, something that did not require interpretation. It was enough to share. They returned quieter but easier, conversation drifting without edge. That night sleep came without calculation. In the morning the trucks were loaded again and pointed east.

Nothing had been solved. Nothing had been escaped. But the pressure on them had eased just enough for their bodies to remember something other than engines, loss, and waiting.

New Command

The pause did not mean the war had slowed. It meant that larger forces had begun to move.

Far above the forward strips and bivouac lines, the campaign was being reorganized by men whose names already carried weight. In the wake of the defeat at Kasserine Pass, General Sir Harold Alexander assumed theater ground direction as commander of the newly formed 18th Army Group, charged with binding the British First Army and the American II Corps into a single front. His concern was not speed but coherence. Armies had to stop moving as separate instruments and begin to strike as one.[2]

Air power was folded into that same logic. Authority flowed downward from Sir Arthur Tedder through Carl Spaatz's Northwest African Air Forces and Jimmy Doolittle's 12th Air Force, compressing fighters and bombers into a single operational rhythm aligned to Alexander's ground plan.[4] Decisions made at desks in Algiers altered the texture of flying hundreds of miles away. Escorts lengthened. Tasking tightened. Sortie timing began to match corps-level movement rather than local convenience.[24]

At General Alexander's urging, Eisenhower relieved Lloyd Fredendall after the disastrous defeat at Kasserine Pass. He placed George S. Patton in provisional command of II Corps, with Omar Bradley beside him as deputy and designated successor.[2,7,8] The change was felt almost immediately at the edges of the system. Where Fredendall had directed from a distance, Patton inserted himself physically.

Camouflage was stripped back from command bunkers and discipline was enforced in the open. Schedules hardened into orders that were written in chalk and erased just as quickly.

Patton walked the lines in polished boots, his ivory-handled .45 riding high on his hip, his voice cutting through wind and dust. "We advance, always."[1] It was less a slogan than a directive, and it carried. The infantry felt it in their forced marches and tightened formations. The artillery felt it in their rates of fire and above them, the 33rd Fighter Group felt the echo in the air. Sorties were drilled as relentlessly as infantry step.

For the pilots, these leaders were distant and omnipresent at the same time. Most would never meet Alexander or Tedder or Spaatz. They would not hear Doolittle speak. Patton might pass through a field without stopping. Yet, those leaders shaped what a morning's flying looked like, how long it lasted, and what it was expected to accomplish.

Watts proved to be an exception. His Air Medal was presented to him in person by Jimmy Doolittle and General Spaatz, the award issued by command of Doolittle and transmitted by Hoyt S. Vandenberg. For a young fighter pilot, it was a rare crossing of levels, a moment when the architects of the air war stepped briefly into the life of a man flying it.[9]

Former Jokers, now scattered through the 58th, 59th, and 60th Fighter Squadrons recognized the pattern instinctively. They had launched once before at the urging of giants they barely knew, pushed off a carrier deck by decisions made far beyond the horizon. That memory remained. It reminded them that history did not move because of them, but that it moved through them.

Between Battles

When the 59th squadron reentered the system, their machines had changed.

The P-40Ls stood lighter on their gear than the aircraft they replaced. Less armor, longer legs. Less punishment absorbed, more distance covered. They were not better in every way, but they could finally remain with the bombers, remain over the fight, remain long enough for repetition to matter. Ground crews learned the new balance points. Pilots learned how far they could lean before the airplane asked to be paid back.[2]

Orders tightened as the equipment improved. Training was no longer something done between missions. It was done because it could not be deferred. Strafing runs were flown until lines formed in the dirt instead of patterns. Dive angles were corrected without discussion. Turnarounds on the ground were shortened. Engines were run hard and shut down hotter than before. Berteaux became a place where nothing was wasted, least of all time.[3]

On 11 March, Watts flew a sea sweep off Tunis and Bizerte, lifting out of Bone before dawn. The coastline slid beneath his wings in a long gray arc, the Mediterranean flat and deceptive, its surface hiding traffic and threat in equal measure. The task was simple and unforgiving: find movement, deny it daylight, return with fuel enough to land. No contact was logged. That absence was its own signal. The planners were at work on something.[1]

The next day that system pulled him back inland.

On 12 March, the 59th flew escort for B-26 Marauders north of Sousse, launching from Berteaux into weather that refused to settle. The formation stretched and compressed as it climbed, engines droning unevenly through turbulence. Somewhere over the route, German fighters broke through. His log records it without embellishment: one B-26 lost to a Me 109. No coordinates. No names. Just the fact of it, written down and left to stand.[1]

The escort returned thinner than it had departed. The ground absorbed the loss without ceremony. Another aircraft was scheduled. Another briefing followed. The day closed without comment, but the lesson remained. The air north of Sousse was no longer permissive. The war was reaching outward from Tunisia's spine, and the fighters were already being asked to hold area that had not yet been taken by ground troops.

By the 3rd week of March 1943, the group was staging forward, shifting toward Sbeitla, with its barley-stubble strips cut into the Tunisian plain.

Across the Atlas ridges, Rommel was massing armor.[2] [5]

Notes

1. Martin Blumenson, Patton: The Man Behind the Legend (New York: Morrow, 1985), 187–191.

2. George F. Howe, Northwest Africa: Seizing the Initiative in the West (Washington, D.C.: Center of Military History, 1957), 146–157.

3. Twelfth Air Force, Combat Diary, March 1943, National Archives and Records Administration [NARA] Record Group 18; see also AAF "Medium Bomber Operations, Tunisia," Intelligence Summary No. 37 (March 1943).

4. Carl A. Spaatz to James H. Doolittle, "Memorandum on Reorganization of Northwest African Air Forces," February 1943, NARA RG 18, AAF Headquarters Files.

5. Williamson Murray and Allan R. Millett, A War to Be Won: Fighting the Second World War (Cambridge, MA: Belknap Press, 2000), 309–312.

6. U.S. Army, After-Action Report, "601st Tank Destroyer Battalion, El Guettar Engagement," March 1943, NARA RG 407, Entry 427.

7. Omar N. Bradley, A Soldier's Story (New York: Henry Holt, 1951), 93–98.

8. B. K. Watts, Pilot Logbook, March 1–31 1943 (Agadir, Berteaux, Sbeitla, El Guettar entries), author's collection; 33rd Fighter Group Combat Digest, April 1984 Reunion booklet.

9. Department of the Army, General Orders No. 32, Section III, Headquarters, Twelfth Air Force, 20 February 1943, awarding the Air Medal to 1st Lt. Blanchard K. Watts (ASN 0661255) for "ten fighter sorties against the enemy" during combat operations in North Africa. The order was issued by command of Major General James H. Doolittle and transmitted (t/) by Brigadier General Hoyt S. Vandenberg, then Chief of Staff, Twelfth Air Force. Watts later annotated his personal copy in pencil, noting "presented by Gen. Spaatz and Doolittle." Author's collection.

CHAPTER 8

AFTERMATH AND ADVANCE

During his first eleven days commanding II Corps, Patton drove east without pause. He dismantled Fredendall's defensive sprawl, collapsed overlapping headquarters, and imposed a tempo that allowed no drift. Artillery drills were hammered into the soldiers until they snapped to cadence. Helmets were painted and weapons were cleaned until they shone. These were not cosmetic acts. They were signals of the new discipline: II Corps would stand, and it would advance.[1] [2]

The 33rd felt the change almost immediately. At Berteaux, tents were struck, propellers lashed to flatbeds, and ammunition and fuel hauled east across the Atlas. Pilots waited through the monotony while new strips at Sbeitla were set up, and their call signs were keyed to a reshaped headquarters. The ground work took days. The intent was immediate.[3]

For the pilots, it was eleven days off the page. No sorties logged. No columns filled. Only preparation, compressed into long hours of waiting and short bursts of movement. For II Corps, it was reorganization in motion, units aligning while already advancing. Everyone understood the test was coming. Beyond the valley, German armor was gathering. Collision was no longer a possibility. It was inevitable.[2] [5]

Opening the Pass

On 17 March 1943, the 1st Infantry Division pushed southeast from Sbeitla toward the El Guettar gap. Italian units of the Superga Division were dug in along the approaches, wire strung low and mines seeded into the fields, but Patton's artillery found the ridges first. Engineers cut lanes under fire. By dawn on the 18th, the Italians had bent back, leaving the valley open and scarred.[2] The air over the valley was already busy, even before Watts returned to the line.

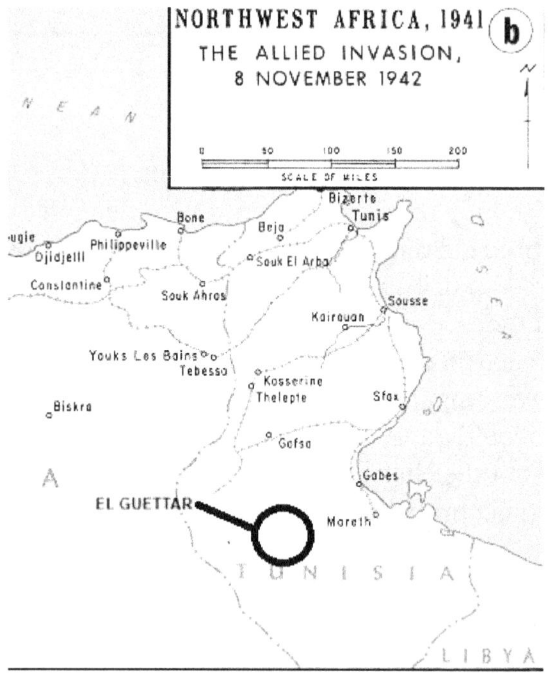

Figure 8-1 El Guettar Tunisia Battleground, Original U.S. Government map, (HMSO, 1956). Public domain. Annotated by the author.

Twelfth Air Force fighters worked ahead of the advance, drawing fire from roads and probing the approaches. Bombers went in under escort, shaping the space the ground forces would soon be forced to occupy. From Berteaux, the 33rd watched the system begin to lean forward.

Figure 8-2 BK Watts's tent in Sbeitla, Tunisia, March 1943 Annotations by BK Watts. (author's collection)

The 59th relocated east as the pass opened. Barley fields at Sbeitla were scraped flat, the PSP was laid, fuel and ammunition hauled in under dust and heat. Orders appeared on boards and were revised before they could settle. Lift times were posted, erased, and posted again. The field filled with the noise of preparation rather than engines.

Watts did not fly during those first days. His log remained blank while the valley was opened and the fight took form below. He walked the strip, studied maps, listened to radio traffic thicken, and learned the names that began to repeat themselves. El Djem.

Kairouan. Mezzouna. They were no longer abstractions. They were routes waiting to be worked.

When he returned to the air on 23 March, the battle for El Guettar was already moving. His first escort and reconnaissance runs entered a fight that was shaped without him, but not without the system he was part of. From the cockpit, the bowl no longer had to be imagined. It was already open in front of them.[4]

The 5th Panzer Army

The counterstrike came on 23 March.

Axis armor moved west through the wadis under dust and smoke, columns fanning where the ground allowed and compressing again as the terrain narrowed. The maneuver carried the unmistakable stamp of earlier campaigns: an armored thrust meant to disrupt, unbalance, and force II Corps to react before it could consolidate. The Fox's way of fighting still lingered in the system, even if the Fox himself was gone.

Erwin Rommel had already departed Tunisia weeks earlier, recalled to Germany and removed from the field. In his place, General Hans-Jürgen von Arnim commanded forces that no longer possessed the freedom, fuel, or air cover that had once made such moves decisive. The German intent remained aggressive, but the means were thinner now, the margins for error were much tighter.[2][5]

II Corps braced while the Twelfth Air Force committed everything it could bring forward on short notice. From their newly cut strips, fighters lifted in steady sequence, the air filling faster than maps could be updated. The system was still aligning, but it no longer hesitated.

That day, Watts logged only a cross-country flight. He relocated east as the front shifted beneath him, flying above a battlefield already in motion. Smoke rose from wadis and ridge lines he would soon learn by heart. Below, the ground war surged ahead of the staging fighter pilots.

The fighting intensified overnight.

On 24 March, BK returned to combat as escort for bombers striking around El Djem. The mission placed him directly into the tightening ground war, covering aircraft committed low and exposed against roads and junctions feeding the armored movement. The airspace was no longer permissive. German flak rose with confidence from stone and scrub, measured and deliberate.

Below him, Shermans burned nose-down in wadis. Infantry clung to ridges no larger than parade fields. German 88s fired from carefully prepared positions, their bursts disciplined rather than speculative. The air escort held, not to seize ground, but to deny the enemy freedom of movement.

By then, the valley had become a bowl. Air and ground pressed against all of its sides at once, each dependent on the other to keep space open. BK's log reduced the day to task and duration. The aircraft that returned carried the rest.

The Axis thrust did not break through. It slowed, then stalled, checked by a system that no longer yielded ground once it had been claimed.

Holding the Bowl

The fight at El Guettar did not resolve in a single blow. It opened with several hits.

On 25 March, Watts flew twice in escort, covering B-25s pushed into the bowl to strike roads and gun positions feeding the German thrust. The bombers went in low, formations stretched by weather and terrain, and the escorts remained with them longer than doctrine preferred. Flak climbed steadily from wadis and junctions, now timed rather than speculative. The valley below looked churned and unstable, dust lifting and settling as armor shifted position.[5]

Later that day, BK flew a reconnaissance run toward Kairouan. The task was observation rather than attack, tracing movement along roads that no longer flowed freely. Trucks halted where they should have moved. Columns bunched, then scattered. The Axis resistance was not collapsing. It was stiffening.

By the morning of 26 March, the air war narrowed to a protective stance.

Watts flew airdrome patrol over the forward strips. This was not hunting. It was denial. Fighters orbited in overlapping patterns, eyes outside the cockpit, waiting for silhouettes to break the horizon. Nothing happening meant everything remained usable.

Below, the PSP matting gleamed where bombs had not yet fallen. Fuel drums were stacked behind berms. Ground crews moved quickly, heads up, tools in hand. A single enemy pass could have undone days of preparation. The air patrol held.

Across the bowl, the ground fight pressed and recoiled in short, violent rhythms. German grenadiers probed at night. Moroccan goumiers worked the flanks by paths no map recorded. Axis artillery spoke in deliberate intervals, American guns answering ridge by ridge. Forward controllers pulled fighters when they could and released them when they had to. The air and ground were no longer separable issues.

By 26 March, however, the German thrust had lost coherence. Columns that had advanced with purpose now burned where they stood. Movement slowed, then stopped. At dawn, smoke hung low and unmoving over the valley. It was not a sign of victory. It was exhaustion.

The bowl held.

Going Forward

For II Corps, the stand at El Guettar altered expectations. American divisions that had broken at Kasserine had now held under pressure, and that knowledge moved faster than reports. It passed through mess tents and maintenance bays, along flight lines and into slit trenches where smoke still clung to the ground. The valley did not feel safer. It felt workable. In Tunisia, that distinction mattered.

What followed was the tightening of a vise:

- Montgomery advanced from the southeast.
- Anderson's British First Army pressed from the north.
- Bradley's II Corps drove east across the central plain.

This was the operational frame now. The Axis absorbed shocks from multiple directions. The front compressed toward Tunis.

The Air War Tightens

Above that movement was the Twelfth Air Force. From Sbeitla and forward fields beyond it, the 33rd FG flew without pause. Watts's log filled with repetition. Routes returned. Targets reappeared. Durations shortened. What appears monotonous on the page reflects escalation in the air.

Morning patrols went first. Midday brought bomber escorts. Late sorties pressed farther east, testing airspace that had once been uncontested. The effect was cumulative. Roads struck repeatedly stopped functioning as roads. Airfields bombed in sequence lost their utility. Each short entry in the log meant fewer vehicles reaching the front and fewer options available to those trying to hold it.[6]

Attrition continued. Aircraft returned with torn control surfaces and streaked coolant. On 4 April, BK flew four sorties. During one of them, his wingman Lt. Mayo was lost and did not return. The loss entered the record quietly and thinned the line again.[7]

By early April, the Joker bond had narrowed further. Carpenter. Cuthbert. French. King. Mayo. Names no longer answered at roll. Survivors carried the change without comment.

Watts kept the ritual. Before takeoff, he saluted the next section rolling out. Decades later, at a 33rd Fighter Group reunion, he would mark a bold J beside every Archer man on the roster. What began as habit endured as record keeping.

Politics in the Rear

Behind the front, authority in North Africa remained unsettled. Admiral Jean-François Darlan, the Vichy commander who had switched sides during TORCH, had been assassinated in December 1942. His death opened a contest between Henri Giraud, supported by the Americans, and Charles de Gaulle, favored by the British. Through early 1943, legitimacy remained unresolved.[8]

In the field, politics appeared only in fragments. French units shared space with American camps. Passwords changed. Mechanics traded parts across Algerian airfields to keep aircraft serviceable. For the men flying and maintaining P-40s, ideology reduced quickly to function.

The Road to Tunis

Watts's log tracked the widening arc of operations.[7]

- 6 April Escort, A-20's, La Fauconnerie, 2:05
- 11 April Bombing & Strafing Enfidaville, 1:15
- 15 April P-40 bombing, southwest of Tunis
- 16 April, fighter sweep, Tunis

Each entry marked another step east. Ferry flights logged simply as X-country bridged the distance between forward strips as II Corps moved through Fondouk and Ousseltia. Every relocation tightened the circle.

El Guettar had not ended the campaign. It had changed its direction. What followed would complete it.

Notes

1. Martin Blumenson, Patton: The Man Behind the Legend (New York: Morrow, 1985), 186–188.
2. George F. Howe, Northwest Africa: Seizing the Initiative in the West (Washington: Center of Military History, 1957), 127–130
3. Howe, Northwest Africa, 138–142.
4. B. H. Liddell Hart, ed., The Rommel Papers (New York: Harcourt, 1953), 442–444.
5. Twelfth Air Force Diary, March 1943, NARA RG 18.
6. Twelfth Air Force Diary, April 1943, NARA RG 18.
7. B. K. Watts, Pilot Flight Log, April 1943 (author's collection).
8. Williamson Murray and Allan R. Millett, A War to Be Won (Cambridge: Belknap Press, 2000), 311.

CHAPTER 9

THE PROVING GROUND

The maps called it Djebel, pronounced "Jebb-il," Tahent,

The soldiers called it Hill 609.

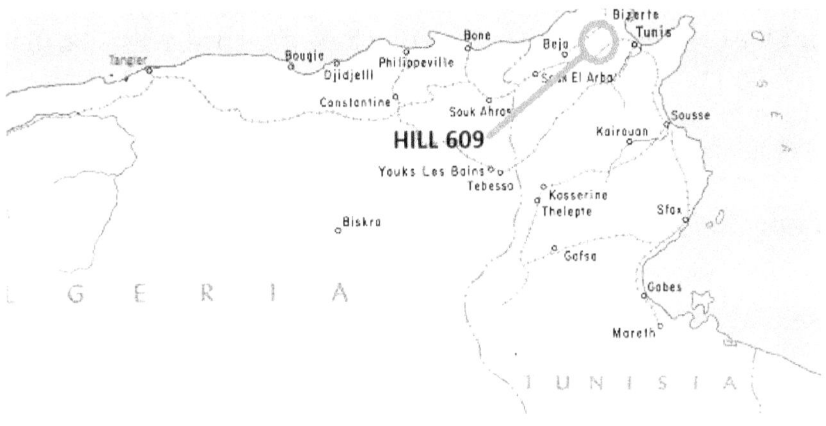

Fig 9-1 Djebel Tahent "Hill 609" located west of Tunis, at 609 meters high, was the key to the final battle in North Africa   Original U.S. Government map (public domain), via Wikimedia Commons, annotated by the author.

It took its name from the elevation figures on the French maps they carried, and from the way it loomed over the approaches below. Djebel Tahent sat at the westernmost hard end of a triangle.

One leg pointed north toward Mateur and the roads running on to Bizerte. The other angled east toward Tebourba and, beyond that, Tunis. From its crown, the German artillery observers stitched those routes together. Any convoy that tried to pass along one arm exposed itself to fire from the others. For the battle-tested 1st Infantry Division, this was not just another tactical obstacle. It was a proving ground.[2]

The rain broke first, thin needles on canvas, a whispering prelude across the bivouac lines. By dawn on 24 April, clouds hung low over Tunisia like a frayed curtain, and the talk around the cook fires had turned to the hills. Men pointed with spoons and cigarettes toward a line of darker humps on the horizon, where Hill 609 stood like a clenched fist over the Mateur corridor. Everyone knew the name now. Djebel Tahent. This hill was the key to all of Tunisia.[2]

On 16 April, General Omar Bradley assumed command of II Corps from George Patton.[4] Bradley, once Patton's understudy, understood the importance of that mound more clearly than most. After Kasserine, Doc Ryder no longer carried a division, but Bradley still trusted him with hard terrain. He sent the order to Doc directly, with a blunt charge: "Get me that hill, and no one will ever again doubt the toughness of your division."[6]

Pont du Fahs

BK's log began to thicken in those days, the ink pressed deeper into the paper as if the pen itself understood what the earth was about to do. The orders were clipped. The timings precise. Between the lines, the stage was being set, a slow tightening of the noose.[1]

Pont du Fahs was the order of the day, not Hill 609. Not yet.

On 24 April, BK flew troop cover, sweeping the southeastern skies while the ground war shifted its weight.[1] Pont du Fahs airfield lay far south of the immediate Hill 609 sector, a separate node in the lattice of air and dust and khaki stretched across Tunisia.

The sortie mattered because airfields always do, and convoys mattered even more. The ridge that would soon take men's names who had not yet felt the full heat of preparation. That day, BK's world was wide country and sky lanes. From the higher view, the theater was aligning itself toward a single point.[1] [2] [3]

Night washed the maps clean. The next morning the ridge felt closer, its angles were sharper and its folds began collecting fog. On 25 April, his flight log stiffened into purpose.[1]

It was a busy day.

The first sortie was cover for the 58th on a bomb run. Escort duty, but with teeth.[1] [3] The P-40s moved low and fast across the corridor, Mateur, Tebourba, and among the arteries that were pumping men and shells into the Hill 609 complex.[2] [3] BK held on station, diving when contrails knotted wrong and when tracers stitched the late morning air into jagged red script.[3] From altitude, the ridge was a geometry problem. On the ground, it was a wager of bodies and nerve.

The second sortie turned aggressive. Bombing gun emplacements northwest of Mateur.[1] [3] The 1st Division (The Big Red One)called it shaping. The pilots knew it as work. Those Axis gun pits did not just command ground, they implanted fear along the routes that mattered. Eventually, the approaches feeding the valley and the heights beyond seemed ready.[2] BK's bombs fell with patient violence, lifting corners of the enemy's curtain and punching daylight into positions that had whispered across the plain for weeks.[1] [2] [3]

The third sortie that day stretched farther. Bombing again northwest of Mateur, then south of Pont du Fahs.[1] [2] [3] By then, the sky felt worked over, thick with smoky residue and intent. BK flew low and steady, stitching roads and junctions into interruption. Dust and flame answered like signals. Upturned wheels spun, then stopped. Convoys scattered. Each pass opened small clefts in enemy movement that the ground forces would widen hours later.[2] [3]

By afternoon, the ground force seemed to lean toward the hill. Maps folded along new creases. Officers traced arcs with pencils, sleeves gray with dust, voices low.[2] Hill 609 had always been the answer to a question everyone knew but rarely said aloud. How do you pry open the door to Mateur? How do you break the German lookouts' hold on the valley? You take the height, or the height takes you.[2]

The men of II Corps felt the air change. Artillery probed the slopes the way a physician tests a wound. Forward observers sent back numbers, and the big guns began to speak in paragraphs instead of sentences.[2] Each shell tested the German resolve, asking whether the enemy would blink.

That evening, the pilots' mess was quieter than usual. BK's flight jacket held the day's cold like a memory. He had flown above Mateur's stubborn geometry, over Tebourba's braided roads, and across fields that looked innocent from altitude but never were.[1] [2] [3] The 24 April mission to Pont du Fahs in support of the British advance still hummed at the edge of BK's thoughts as a reminder that the campaign was a web, not a line. But, it was the work on April 25th that settled in his bones. Those sorties still felt close to the ridge, close enough to hear the hill breathe.[1,2,3]

On the ground, companies rehearsed their own arithmetic. Ammunition stacked like cordwood. Bandoliers were checked twice.[2] The chaplain walked the lines with an unhurried pace. Stories began and ended without middles. A private stared at the hill until his eyes watered, laughed at himself, then stared again. It was a ritual, as if memorizing the contours could will a path up through the stone.[2]

Hospital tents brushed the wind behind them. Ahead, outpost fires burned in thin pockets of flame.[2] Between those two points, cause and consequence, the American line settled its weight and gathered the long breath men take when they believe the next one may not come.

By twilight on the 26th, the sky closed like a lid. BK closed his log and ran a hand over its worn leather, feeling the faint ridges left by pressed ink.[1] The ridge beyond showed in silhouette, black on black, promising a fight that would invent its own language of scarp and spur, wire and stone, with grenades breaking in rock hollows. Tomorrow or the next day, the 34th Division would put their shoulders to that task. The 1st would feel the push spill through the valleys.[2]

Already, the hill had been disturbed. German observers blinked behind their glasses. The routes feeding the position felt the hot breath of Allied engines and the blunt punctuation of its bombs.[1][2][3]

In the morning, II Corps would try to take the hill, regardless.[2]

The Assault

On 27 April 1943, the 34th Infantry Division moved forward with armor and artillery.[2] On the slopes waited hardened defenders, Fallschirmjäger and veterans of the 334th Division, dug in for weeks with every wadi sighted and every ravine pre-registered.[2]

Mortars rained into gullies where Allied riflemen clawed upward. Shermans shuddered under precisely observed fire. American gunners tried to smother the summit with 155s, and every silence broke again under answering German rounds.

Fighter sweeps overhead suppressed German reconnaissance, but air to ground coordination remained crude. The Twelfth Air Force experimented with Rover Joe missions, radioing fighters toward hot targets.[3] Watts logged steady work, patrols, escorts, armed reconnaissance. Patrol escort 2:05, he wrote on 27 April. What he saw was smoke crawling the ridgeline like wool snagged on wire, artillery haze breaking the hill into glare and shadow.[1][3]

BK's map softened at the creases from folding and unfolding, rehearsal for the violence was traced in pencil before it was written in blood. His flight log became the score. Each entry precise. Each one tracking the tempo as Bradley's II Corps campaign tightened from ridge to shore.[1][2][3]

A Pilot's Ledger

On 28 April, BK flew again. His first sortie lasted 1:15, bombing gun pits near Mateur and Tebourba that had whispered at American lines for days.[1][2][3] His second sortie that day escorted the 58th on the same run. Shaping in its plainest form. Strangle the Axis roads that were feeding Hill 609.[1][2][3]

Below, the forward companies felt the change as pressure in the air. Somewhere on the slope, a German observer blinked at sudden daylight where his position had been.[2]

By his fourth sortie in three days, BK hunted the arteries of the plain, Mateur again, then a low sweep south toward Pont du Fahs.[1][2][3] He rearmed at Ebba Ksour in a P-40L, dust dulling the olive drab, with Sgt Reed's braided pull still tugging at the cockpit hinge. Bomb clusters hung beneath the wings like blunt promises.[1]

Roads, junctions, timber approaches. Trucks stalled. Wheels spun and stopped. Drivers scattered into scrub. Each pass BK made lanced another lane through enemy movement, seams that the infantry would widen by nightfall.[2][3]

In late April, the arc curved inward. BK worked south of Mateur, then toward Tunis.[1][2][3] From altitude, the campaign revealed itself as two struggles bound together: the intimate brawl on the hills and roads, and the larger contest of the ports and airfields. BK carried both simultaneously and returned to base with the sense that both ridge and city were part of a single breath the campaign was learning to control.[1][2][3]

The ground absorbed each note. Maps folded differently now. Artillery spoke in paragraphs.[2] The hill began to give, not with collapse but as a loosening, as if the mountain itself had exhaled. Still, the Allied pressure held.[2]

By 30 April, it seemed the planners knew something. BK was already working south of Mateur, then back toward Tunis. The arc curved forward and then inward.[1][3] Down there, he traced the seams that bound the ridge to its lifelines, punching at the last threads feeding the fight on Hill 609. Then his nose turned east and the horizon widened toward Tunis, the city that loomed as both prize and problem. From sky to slope and then to the road, the Allied pressure held.[1][3]

Breaking the Grip

The struggle for Hill 609 lasted four days. Infantry of the 34th Division clawed upward with bayonets fixed. Artillery thundered every quarter hour, fulfilling Doc Ryder's pledge that he made weeks earlier.[5][6]

Air cover prevented Axis daylight resupply. Their convoys burned along approach roads.[2][3][5] Still, every yard was taken by men on foot.

On 1 May, under cover of smoke, German forces withdrew. At dawn, American troops stood on the bare summit. The cost was heavy, nearly 2,500 casualties in the 34th Division alone, but the victory was real.[5]

Word spread faster than memos. If Kasserine had been humiliation, and El Guettar justification, then Hill 609 became vindication. For the first time, American infantry, artillery, and air power had broken a fortified German position by combined weight and will.[25]

Toward Tunis

With Hill 609 taken, the Axis grip on central Tunisia cracked. Bradley's columns streamed east. Anderson pressed from the north. Montgomery closed from the coast.[2][5]

The coil, wound since November, finally loosened. Tunis lay days away. For Watts, it meant the next flight. Another sortie. History grinding forward, one log entry at a time.[1][2][3]

Notes

1. Blanchard King Watts, Pilot Flight Logbook (April 1943, author's collection).
2. George F. Howe, *Northwest Africa: Seizing the Initiative in the West* (Washington: CMH, 1957), 151–54.
3. U.S. Army Air Forces, *Combat Operations Diary, April 1943* (NARA RG 18), 72–73.
4. Carlo D'Este, *Patton: A Genius for War* (New York: HarperCollins, 1995), 520.
5. George F. Howe, *Northwest Africa*, 154–56; Williamson Murray and Allan R. Millett, *A War to Be Won: Fighting the Second World War* (Cambridge: Belknap, 2000), 313.
6. Bradley, Omar. A General's Life. New York: Simon & Schuster, 1983, p. 156.

CHAPTER 10

THE FALL OF TUNIS AND BIZERTE

By early May 1943, Tunisia had become a vast funnel. Every remaining Axis formation, especially the remnants of the Afrika Korps, had been forced into its narrowing neck. Von Arnim's Fifth Panzer Army, already battered since Kasserine, now consisted of broken battalions and exhausted units fighting for space and for their very lives. To the north lay the twin ports of Tunis and Bizerte, and beyond them only the Mediterranean. Eastward, Allied bombers flew almost unopposed. Westward, Omar Bradley's II Corps pressed forward through olive-clad hills. In the south, Montgomery's Eighth Army advanced with the patience of a man winding a clock.[2]

BK felt the change in the air before the boots secured it on the ground. Axis convoys below grew ragged and Luftwaffe escorts appeared only sporadically. Axis flak still jabbed upward, but its sting had dulled. His log showed no sorties for the first four days of May.[1] Even that pause carried meaning. Below, trucks moved differently now. Axis hulks smoldered at crossroads and the Allied columns surged forward with fewer halts and the skies leaned decisively toward the star-and-bar insignia of the Allied air forces.

The Allied Jaws Close

By early May, the campaign had taken on the clear geometry of entrapment. Along the southern coast, General Bernard Montgomery's Eighth Army hammered the Enfidaville Line. His gunners cracked one strongpoint after another, concrete and wire splintering in the cadence of his method: fire, move, consolidate and fire again.[2]

To the west, Bradley drove II Corps out of the hills above Mateur. Shermans growled through olive groves. Tank destroyers ground their way over stony ridges. Infantry clawed through ravines by hand and bayonet. For II Corps, pauses were sin. Orders drove the divisions forward day and night.[3]

From the north, General Kenneth Anderson's First Army pressed down the Medjerda Valley, squeezing toward Tunis. One arrow pointed south, another east, another north.

Together these massive armies formed a closing vise: Montgomery methodical in the south, Anderson pressing from the north, and II Corps ramming forward from the west. The Axis pocket shrank by the day, driven back toward Tunis and Bizerte, with no road left but surrender or collapse into the sea.

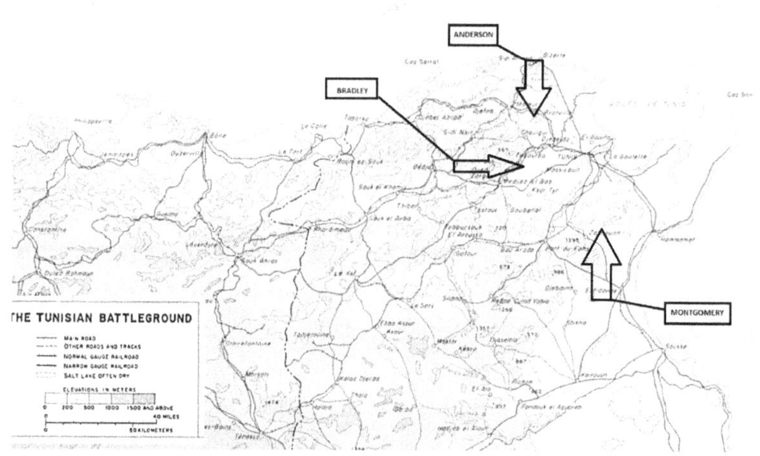

Fig 10-1 The Battle of Tunis Tunisia Battleground, 1942 (Map VI). Drawn by D. Holmes for the UK official history The Mediterranean and Middle East, Vol. IV: The Destruction of the Axis Forces in Africa (London: HMSO, 1956). Annotations by Author. Public domain

The View from the Cockpit

The Luftwaffe flashed its last defiance on 5 May, with Bf 109s slashing twice through Allied formations and Ju 52 transports risking impossible daylight re-supply runs. On one such mission near Zaghouan, Watts' log recorded: "One burst: fire aft, transport ditched in wadi."[1]

Meanwhile, the Twelfth Air Force hammered the shrinking Axis pocket with nearly 1,500 sorties that week.[4] El Aouina and La Marsa airfields became cratered graveyards. Any Axis aircraft attempting escape risked being fireballed over Tunis Bay.

Inside that deluge, Watts' log captured the muscle-ache tempo of a single pilot. Between 29 April and 11 May, he flew seventeen combat sorties, many doubled, one day flying three in succession.[7]

His logbook charted the collapse with blunt economy:

- 5 May: Medjez el Bab sortie; cover for ship bombing in the bay off Tunis.
- 6 May: Bombing north of Pont du Fahs.
- 7 May: Bombing Tunis airfield.
- 8 May: Bombing shipping off Metline; bombing airfield at Menzel Temime.
- 9 May: Two sorties bombing traffic north of Soliman.
- 10 May: Two sorties bombing gun emplacements at the end of Cape Bon.
- 11 May: Two fighter sweeps over Cape Bon.

After the May 11 entry he wrote, almost with relief, "Campaign Ended."[1]

The theater communiqués counted thousands of sorties per week. In Watts' hand, the end of the African campaign measured in a few grams of ink. Sorties compressed into four words written down that closed a seven-month long ordeal.

Cities Fall

On 6 May, the British 6th Armoured Division entered Tunis, with flag-waving civilians pouring out from the alleys. That same day Bradley's II Corps drove into Bizerte, seizing its naval base in the name of vindication.[3]

Below, depots burned, bells tolled, flags streaked the windows. That evening brought a haunting tableaux: In the square, a weary German POW with an accordion, playing for his captors. Around him, American and British troops sat in circles on crates under the trees, silent, half smiling, too exhausted to jeer. Humanity had returned in the strangest disguise. In contrast, on the outskirts of Tunis, BK photographed a German military graveyard.

From 7-13 May, the unraveling became surrender. Axis troops by the tens of thousands filed into captivity. Eisenhower's bulletin read: "The Battle of Tunisia has ended."[5] For Watts, it ended already: 11 May, Cape Bon. His constant sorties gave way to quiet tents and the dull expectancy of what came next.

Fig 10-2German POW playing accordian for the Allied troops in Tunisia Annotations by BK Watts. (Author's Collection)

By 13 May 1943, the reckoning was staggering: more than 250,000 Axis troops, including twelve generals, surrendered. This was greater even than the count at Stalingrad.[6] An entire Axis army group was erased in a fortnight.

For Montgomery it vindicated patience. For Omar Bradley

it proved Americans could do more than endure, it showed they could overwhelm. The campaign in North Africa had ended.

As for George Patton...well...he wasn't quite finished.

Fig 10-3 German graveyard outside Tunis in May 1943 (Author's Collection)

Notes

1. Blanchard K. Watts, Pilot Flight Logbook (Apr–May 1943, Author's collection).
2. George F. Howe, *Northwest Africa: Seizing the Initiative in the West* (Washington: CMH, 1957), 161–65.
3. Carlo D'Este, *Patton: A Genius for War* (New York: HarperCollins, 1995), 530–34.
4. U.S. Army Air Forces, *Combat Operations Diary*, May 1943 (NARA RG 18), 88–91.
5. Dwight D. Eisenhower, *Crusade in Europe* (New York: Doubleday, 1948), 177.
6. Williamson Murray & Allan R. Millett, *A War to Be Won* (Cambridge: Belknap, 2000), 315–17.

7. Blanchard K. Watts, Pilot Flight Logbook (Apr–May 1943, Author's collection).

PART III
CIAO ITALIA

By summer 1943, the campaign in North Africa was finished. The Mediterranean lanes were open. Across Tunisian and Libyan harbors, an Allied invasion fleet gathered, bows pointed east.

The next target was Sicily, long described as the soft underbelly of Europe, though its beaches bristled and its hills stood hard with German fortifications. On the sides of trucks and in camp banter, one word surfaced again and again, equal parts farewell and promise: Ciao.

One island fortress still stood in the way: Pantelleria.

Goodbye Africa. Hello Italy.

CHAPTER 11
ACROSS THE MED

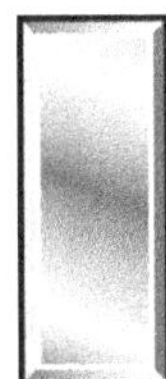

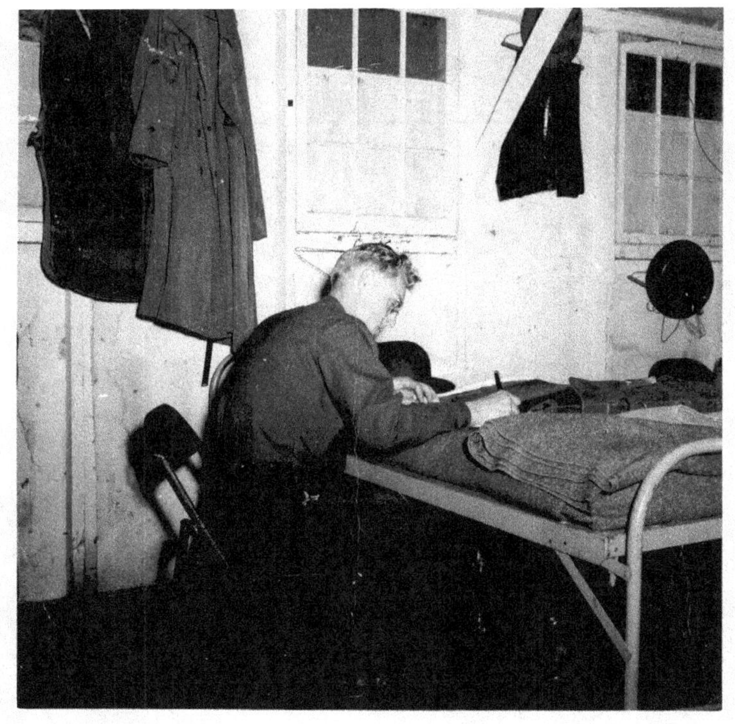

Fig 11-1 BK Watts working on flight logs prior to the invasion of Pantelleria July 1943. (Author's Collection)

The smell of Tunisia still clung to everything. Diesel. Dust and the tang of burned convoys on valley roads. For months the horizon had meant the Atlas mountain ridges and lush olive groves. Now, in June 1943, the pilots of the 33rd Fighter Group stared east across the Mediterranean, wondering how to write the next line of a logbook when a sea interrupts the page.

The last firm line of the North Africa campaign sat alone in BK's hand.

"11 May. Ftr Sweep Cape Bon. Campaign Ended."[1]

There was no flourish. No summary. Just a period. The war did not pause to acknowledge it. The very next entries moved forward, not north or west, but across water. When Allied planning maps came down on tables in Casablanca and Algiers, one word rose above the rest.

HUSKY.

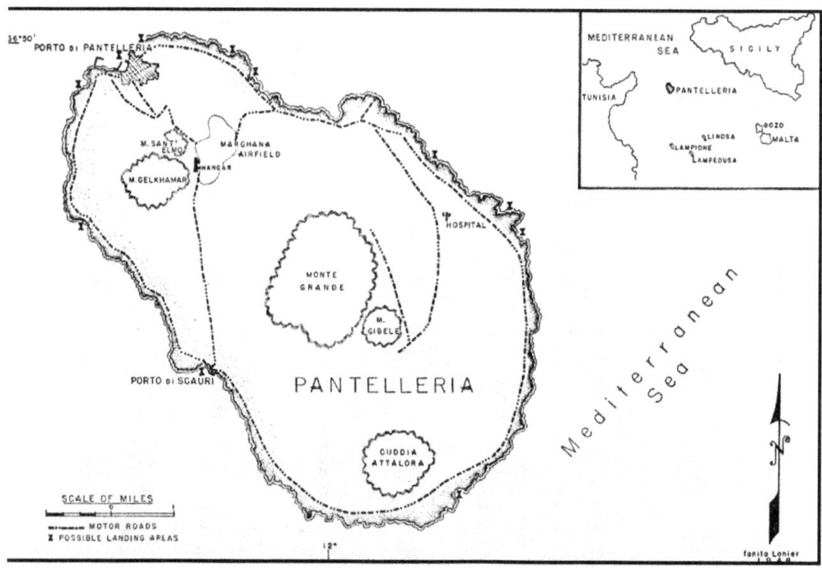

Fig 11-2 Map showing Pantelleria and the Pelagie Islands, including Lampedusa, in the Mediterranean (US Army Air Forces map) Public Domain.

The Forgotten Battle

Before Sicily, however, there was Pantelleria.

It was a volcanic island bristling with Italian guns and German detachments, carved with galleries, caves, and reinforced pits. It also, unfortunately for its temporary residents, sat squarely between Tunisia and Sicily, too valuable to bypass and too close to ignore. It seemed that the Axis had arrived at the very same conclusion much earlier in the war.

From 24 May through 3 June, BK's logbook became a looming list of hammer blows.

Day after day, the same entry repeated in different ink weights. Bombing gun emplacements. Bombing coastal batteries. Bombing ferries and shoreline positions.[1] There was no variation because variation was not the point. The island was being reduced systematically, position by position, air bombardment was substituted for infantry.

Only after 3 June did the pattern loosen. Bombing gave way to fighter sweeps and escort duty. A-20 Havocs. B-25 Mitchells. Patrols over water.[1] The shift was quiet but telling. Pantelleria was no longer being broken. It was now being contained.

Throughout this phase, the 33rd still flew from Tunisia out of strips at Menzel Temime. African dust on boots and tires. The larger war had not crossed the water yet, even if the targets had. That would not change until 24 June.

Pantelleria shook under the weight of it all the same.

On some days, more than five hundred Allied sorties struck the island.[2] Antennas vanished. Caves collapsed. The airstrip cratered until landing there became impossible. Italian morale failed under bombardment that never escalated and never stopped. There was no invasion force assembling offshore.

On 11 June, the Axis garrison raised the white flag.[2]

Not one Allied soldier stormed the island. Pantelleria fell without infantry setting foot on its rock.

Proof of Concept

For Eisenhower's headquarters, Pantelleria was validation. He later called it a useful preliminary to HUSKY.[3]

Tedder and Spaatz treated it as something more specific. Pantelleria was an unsinkable aircraft carrier. From its strip, fighters could range over convoys, beaches, and Axis airfields across western Sicily. Air superiority could be staged forward before the first landing craft touched sand. Omar Bradley understood it the same way. The island was taken not for its territory, but for its position.[4]

There was deception layered into the decision as well. Fighters shifting east could be explained as convoy cover. Radio traffic and double agents pointed Axis attention toward Greece and Sardinia. Pantelleria absorbed sorties while the larger plan matured behind it.[5]

Behind the scenes, however, unity frayed. General Bradley later wrote that the air plan for HUSKY remained opaque to ground commanders. Naval schedules advanced on their own logic. Fighter cover existed, but coordination lagged. His frustration centered on Tedder, whose intentions were clear at the operational level but indistinct where soldiers would depend on them.[4]

Bradley's concern was not academic.

An Active Island

When the 33rd finally moved forward on 24 June, Pantelleria did not feel like a conquered place. It felt charged, as if the island were holding something in reserve. Loaded was the word that fit. Loaded with attention. Loaded with consequence.

From 25 June through the first days of July, BK's log thickened again. Alerts appeared where none had been before. Scramble calls broke the routine. Patrols launched, were cut short, then launched again.[1] The pilots learned to stay close to their aircraft, helmets never far from reach. Pantelleria sat exposed between two continents, its radar sweeping empty water that might not stay empty for long. The sea looked calm. That did not reassure anyone.

On 28 June, BK logged a sighting that carried weight far beyond its few words: Enemy parachutist.[1]

There was no follow-up notation. No clarification. None was required. Someone had been watching the island closely enough to put a man in the air over it. Whether he came down alive or not mattered less than the fact that he had been there at all.

84

On 3 July, BK escorted a flight of A-20s north toward Sciacca.[1] On paper, it read like another routine escort, one more line in a month crowded with them. In the air, it did not feel routine. One aircraft failed to return. Deal's name appears only once in the log, written without emphasis, without annotation. The absence that followed was louder than any after-action report.

After that, time seemed to tighten. Briefings shortened. Launches came closer together. The island felt as if it were drawing breath.

The next entries in the log would no longer be preparation. They would be execution.

Notes

1. Blanchard K. Watts, Pilot Flight Logbook, May–July 1943, author's collection.
2. U.S. Army Air Forces, Combat Operations Diary, May–June 1943, NARA RG 18.
3. Dwight D. Eisenhower, Crusade in Europe (New York: Doubleday, 1948), 182.
4. Omar N. Bradley and Clay Blair, A General's Life (New York: Simon and Schuster, 1983), 144–147, 177.
5. George F. Howe, Northwest Africa: Seizing the Initiative in the West (Washington DC: Center of Military History, 1957), 260–266; Samuel Eliot Morison, History of United States Naval Operations in World War II, vol. 9 (Boston: Little Brown, 1954), 12–15.

Chapter 12

Operation HUSKY

On the night of 9 July 1943, the Mediterranean flickered like a second sky turned upside down. Navigation lamps blinked in coded rhythms. Signal lights pulsed and vanished. Beneath them, phosphorescent wakes unraveled behind hulls and landing craft, ghostly threads stitched across black water. From altitude it looked alive, as if the sea itself were moving with intent.

More than 2,500 ships were underway, decks crammed with men, tanks, and crates tagged for beaches few Americans had known by name until that week. Licata. Gela. Scoglitti.[1]

From the volcanic strip on Pantelleria, captured only weeks earlier, the P-40 squadrons of the 33rd FG clawed into the night from hastily patched runways. The island had become that unsinkable aircraft carrier the planners hoped for, staging fighters forward as the invasion unfolded. Each sortie followed the same arc. Pantelleria to Licata. Convoy cover in rotations that ran long past two hours. Then, back across the water to refuel and go again. BK's tasking did not begin with the spectacle in the dark.

It began with daylight discipline.

On 10 July, flying out of Pantelleria, Watts was assigned cover for the Licata landings. The sortie ran long, more than two hours over the beaches and approach lanes, orbit after orbit while the first waves came ashore in Sicily. Landing craft pressed toward sand. Ships held station. Smoke rose unevenly from the shoreline. His job was simple and unforgiving. Stay overhead. Keep the sky empty.[1]

The following day, 11 July, the assignment repeated.

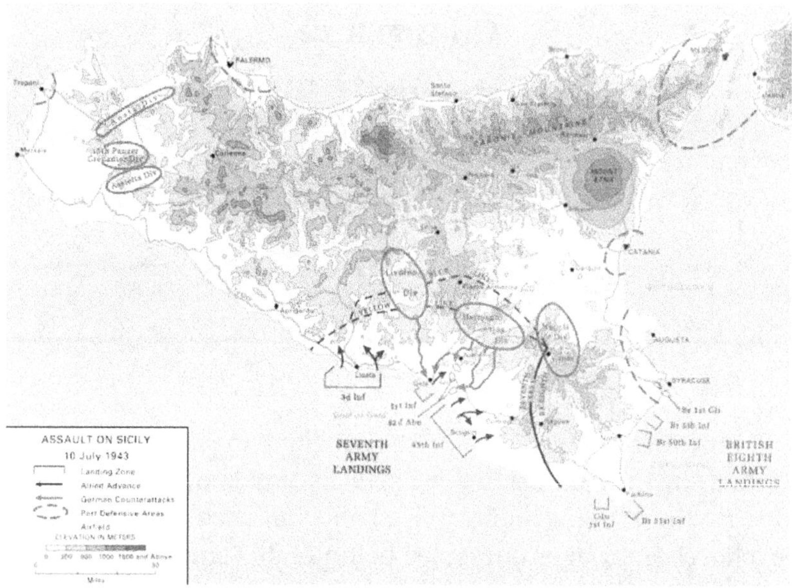

Fig 12-1 United States Army Center of Military History. "Map 1, Sicily Campaign" (from CMH Pub 72-16: Sicily and the Surrender of Italy) Washington, DC: U.S. Army Center of Military History. http://www.history.army.mil/brochures/72-16/map1.JPG (accessed September 24, 2025). Public domain.

"Cover landing, Licata."[1]

Same track. Same altitude. Same patience. The mandate did not change. The beaches were still fragile, the lodgment still thin. Whatever came out of the air had to be seen early and met immediately.

From 12 through 14 July, the mission shifted but did not loosen. Watts flew cover ships at Licata, long patrols over the anchorage and seaward approaches, each sortie stretching well beyond two hours. The beaches had become roads. The roads had become supply lines. The fleet below remained exposed. Luftwaffe appearances were rare, but the absence itself had to be enforced.[1]

On 16 July, his log records a change that mattered.

"Cover C-47s. Moved to Sicily."[1]

The air war followed the army inland. Transport aircraft threaded the sky, hauling men and materiel forward as the beachhead thickened into occupation. That same day, the tasking turned outward again.

The sea would demand attention.

The Search

Later in the day on the 16th, Lt. Howard Young, a pilot from the 59th, had been hit by flak over the sea and forced to ditch. His P-40 vanished into the Med.

BK was flight leader that day. When the call came, it was not abstract. This was not convoy geometry or patrol timing. This was one of their own.

The logbook records it without emphasis:

"Sea search for Young."[1]

What the line does not show is the duration of time Young was in the water.

The search ran hour after hour. Expanding squares. Then tightening again. Whitecaps by day. Black glass by night. Every glint might have been wreckage. Every shadow could have been a raft.

BK stayed low, scanning until distance lost meaning and the compass became the only fixed thing left.

On 17 July, he finally spotted the pilot.

A shape wrong against the water. Too deliberate to be driftwood.

He dropped lower, confirmed it, and stayed with the position while he vectored rescue aircraft in. Seas were rough. Bearings had to be checked again and again.

Howard Young came out of the water alive.

BK's log gives the moment no flourish. No underline. No note beyond continuation of tasking. But within the squadron, the story endured.

He went looking for one of his own...And found him.

The invasion, however, did not pause. The beachhead expanded. Sorties resumed. But for two days, amid the largest amphibious operation in history, the war narrowed to a single life on open water and fellow pilots who refused to turn away.

Licata and the Gulf of Gela

BK's station was Licata. The 59th settled into that role once the beachhead formed, protecting Patton's western flank as men and matériel poured ashore. It echoed TORCH eight months earlier. Casablanca then. Licata now. Hold the air while the lodgment on the beach took shape.

To the east, at the Gulf of Gela, the weight of the invasion fell the hardest. That harbor, with its road north and symbolic gravity, was Patton's focus. Here the infamous Hermann Göring Division counterattacked with armor and infantry driving into beaches that were still cluttered with half-landed Shermans and paratroopers scattered by storms. Despite Allied deception pointing toward Greece and Sardinia, Kesselring had not been fooled. He had positioned his best division for exactly this fight.[3] [4]

Destroyers fired broadside into advancing Axis armor. Smoke sheets rolled across the sand. Headlines later crowned Gela the crucible of HUSKY.

From his cockpit above Licata, BK saw Gela only in glimpses. Smoke columns. Bursts along the coast. The sense of motion everywhere. It was TORCH again, only larger, louder, and much less forgiving.

His log does not mention Gela by name until later in July. His task was narrower: Keep Licata uneventful.[2]

Onshore, the invasion diverged. Montgomery's Eighth Army struck at Pachino in the southeast, brushing aside Italian defenders who fired briefly and dissolved. Patton's Seventh Army landed at Gela, Licata, and Scoglitti. Gela bore the counterattacks for the German divisions dug into the hills above the beaches. Licata held firm under air cover, much of it invisible, built from overlapping fighter patrols like those flown by the 59th.[1][3][4]

For Patton, it was prestige. For BK, it was rotation after rotation, convoy cover stitched end to end until Sicily itself became the operating base.

Inland Missions

After 17 July, the air war shifted inland.

BK's log turns from "Cover Licata" to escort duties for C-47 transports, streams of aircraft lifting supplies and men into Sicily's interior. By 19 July, missions ranged as far as Marsala, then deeper still. Strafing docks. Reconnaissance runs shadowing Patton's columns as they pivoted west and north. Enna followed. Villarosa. Then, maritime strikes off Catania by month's end.

These were high-tempo days. Often two sorties daily. Escort and reconnaissance back to back.

By 31 July, the Warhawks themselves showed the strain. Aircraft were worn out. Pilots were visibly thinner. The squadron crossed back to Pantelleria, not to retreat, but to refit. Palermo had fallen. Messina loomed in the near future. The ground campaign pressed on.

The 33rd FG as Patton's Wing

By the time the war reached Sicily, the 33rd Fighter Group under Colonel Momyer had become, in practice, if not formal orders, Patton's air cover.

On paper, the 33rd was a Twelfth Air Force fighter group. On the ground, the pattern looked different. When Patton pushed, the 33rd pushed. When his columns clawed along the north-coast road or crossed the island's broken spine, Warhawks were usually strung out overhead, angling back and forth along the route.

From the cockpit, BK felt it as overlapping work rather than a single directive. One hour strafing a junction German traffic might use to cut Patton's advance. Then, the next hour they were bombing a gun position shelling an armored column. Later, orbiting ahead of the marching battalions, they watched for dust plumes that meant trouble.

Ground troops learned the sound and shape of the P-40s. More than once, infantry and tank crews told visiting pilots that "our fighters" had been overhead when they needed them most. In Sicily, that usually meant Momyer's 33rd.

If Patton's legend was built on speed and refusal to stop, then a quiet part of that legend rode on the backs of the fighter pilots who cleared his roads and covered his flanks. For much of the campaign, that meant the P-40s of the 33rd Fighter Group. In a very real sense, the 33rd FG was Patton's wing.[6]

Notes

1. Blanchard K. Watts, Pilot Flight Logbook, July 1943, author's collection.
2. George F. Howe, Northwest Africa: Seizing the Initiative in the West (Washington, DC: Center of Military History, 1957), 175–178.
3. Carlo D'Este, Patton: A Genius for War (New York: Harper-Collins, 1995), 541–555.
4. Samuel Eliot Morison, History of United States Naval Operations in World War II, vol. 9, Sicily–Salerno–Anzio (Boston: Little, Brown, 1954), 18–20.
5. Dwight D. Eisenhower, Crusade in Europe (New York: Doubleday, 1948), 182.
6. George F. Howe, Northwest Africa: Seizing the Initiative in the West; Martin Blumenson, Patton: The Man Behind the Legend (New York: William Morrow, 1985).

CHAPTER 13

THE RACE TO MESSINA

By the closing weeks of July 1943, Sicily was no longer a tenuous bridgehead.

It had become a racetrack.

Across the island, two massive Allied armies strained not only against the Germans but against each other. On the right-hand side of the map, Montgomery's Eighth Army, battered but relentless, pressed northward along the lava-scarred Catania plain. On the left, Patton's Seventh Army had swung west into Palermo, then turned right and drove along the island's northern coast toward Messina, hugging cliffs and coves as it chased the Strait.

They may have fought under different banners, but they shared the same sky.

The 59th circled in that sky. Their work was rarely glamorous. Their victories were rarely announced. For them, the "race" looked like long turns in orbit, eyes aching against the Sicilian haze, radios rasping with clipped codes, fuel gauges were being watched as closely as the horizon.

Lessons Learned

Patton's movement made headlines, driven by an army that moved fast and refused to pause. Above it hung that fighter net, largely invisible, that reduced risk, dulled counterstrikes, and made German interference costly when it appeared.

That net, however, did not arrive fully formed.

At the very opening of HUSKY, the system tore into itself. On the night of 11 July 1943, Allied C-47s carrying paratroopers of the 82nd Airborne flew into a storm of their own flak over the invasion beaches. Ship gunners and shore batteries, shaken by Luftwaffe raids, opened fire. More than twenty US AAF transports and tow aircraft were destroyed or crippled. Over a hundred Americans became casualties to friendly fire. A night meant to open the campaign instead exposed the consequences of misalignment.[9]

For the men flying cover, the lesson arrived indirectly. Watts's log for 10 through 14 July records repeated sorties "covering ships off Licata," steady patrols above the same anchorage whose guns had torn into American aircraft days earlier. From the cockpit, the orders did not explain. They simply tasked. Protect the fleet and keep the sky orderly.[19]

The correction in air communication strategy came quickly. XII Air Support Command tightened recognition procedures, redrew flak corridors, and they imposed stricter altitude and timing discipline. By the time Patton's columns were racing along the island's northern rim, those lessons were baked into every briefing. The air and ground systems were no longer learning at each other's expense.[9]

Gela at Last

On 22 July, Watts finally circled Gela itself. From altitude, the scars still showed. Half-sunken landing craft lay like ribs in the surf. Burned tanks lined the inland roads. Smoke drifted from fields where the Hermann Göring Division had fought to a standstill weeks earlier.[2]

He traced it once, then wrote it down without flourish.

"22 Jul. Patrol, 1:55 Gela."[1]

That same day Patton's vanguard rolled into Palermo to cheering crowds, wine and bread pressed into hands, flags shaken from balconies. Watts banked over wreckage while Patton smiled for cameras. Both moments belonged to the same campaign. They did not feel related from where either man stood.

Joy Ride

Three days later came an interlude that felt almost unreal.

On 25 July, the squadron discovered an abandoned Caproni Ca.100 trainer at Licata, its wooden fuselage polished by neglect. Watts coaxed the engine to life and waved his mechanic, Ervin Hunstead, aboard while Sgt. Reed urged them on.

They lifted slowly into a softer sky. Hunstead, usually bent beneath broken engines and fuel drums, saw the island from above. Convoys nudging crude jetties. Columns trailing dust into hills. Bivouacs scattered among olive groves.

Watts dipped his wings.

For a few minutes, the war loosened its grip. It was not forgotten, It was just set aside for the afternoon.[1]

Double Kill Day

The pause did not last.

On the morning of 8 August 1943, General Lucian Truscott's Rangers executed an end-run, slipping ashore near the San Fratello Ridge. The landing sites lay along Sicily's northern edge, between the steep hills and the narrow beaches that left little room to maneuver. The 59th orbited above from Licata, guarding the thin sea lane that fed the ground force's advance.[4]

Out of the haze came what every pilot waited for: Messerschmitts driving in on the shipping.

Watts rolled his P-40 toward them and pushed the throttle forward. One Me 109 flashed belly-up in his gunsight. Six Browning guns stitched a line from spinner to tail. The aircraft fell burning into the Tyrrhenian Sea.

The sortie ended. The war did not.

That afternoon, still flying from Licata, Watts launched again, this time covering the landings near Capo d'Orlando, farther east along the same northern coast where Patton's columns were pressing toward the strait. These attackers were Focke-Wulf 190s, heavier and faster than their Messerschmitt counterparts. Watts ran one down and hammered it until it plowed into scrub beyond the beach. He damaged another Fw 190 badly enough to earn a probable kill.

In the logbook, the entries stayed sparse. In pencil, he marked the margins with two small red swastikas.

- 8 Aug. Intercept, 1:50. 1 Me 109 destroyed
- 8 Aug. Cover landings Capo D'Orlando. 1 Fw 190 destroyed, 1 Fw 190 probable[1]

Two kills in one day was a rarity. It carried weight beyond the ink.

To the squadron, it confirmed what had been building quietly for months. Watts was no longer just the pilot who brought others home. He was a man who saw clearly in chaos and acted without hesitation. That kind of steadiness did not go unnoticed.

Messina

While fighters orbited overhead, Montgomery battered forward on the right-hand side of the island, trading yards along the Catania plain. Patton sprinted on the left, column by column, leap by leap along the northern coast.

On 17 August, American scouts rattled into Messina. They found a city hollowed by shellfire, smoke drifting over the waterfront and the church bells were silent. Across the strait, German rearguards ferried men into mainland Italy under smoke screens.

Seventh Army buglers sounded the occupation. Hours later, Montgomery arrived.

History would record it cleanly: Sicily was secured.

For the men who flew and marched along its length, the liberation of Sicily felt earned the hard way. Logged in pencil but paid for in hours, fuel, and nerve. For Watts, the campaign closed the way so many others had: Escort. Patrol. Bombing, Ferry

Two red marks in the margin were the only exceptions.

Messina was not just the end of an island.

It was the threshold of Europe.

Fig13-2 United States Army. "Generals Montgomery and Patton in Sicily," photograph. Wikimedia Commons. https://commons.wikimedia.org/wiki/File:Generals_Montgomery_And_Patton_in_Sicily.jpg (accessed September 24, 2025). Public domain.

Notes

1. Blanchard K. Watts, Pilot Flight Logbook, July–August 1943, author's collection. Entries for 20–31 July and 8 August 1943 document escort, patrol, bombing, ferry, and intercept sorties, including two confirmed aerial victories and one probable on 8 August, marked in red pencil in the margin of the logbook.

2. George F. Howe, Northwest Africa: Seizing the Initiative in the West (Washington: Center of Military History, 1957), 182–184. For the transition from the Sicilian landings to inland operations and continued air activity over Gela and

the southern coast.

3. Carlo D'Este, Patton: A Genius for War (New York: HarperCollins, 1995), 556–561.

4. For Patton's movements across western and northern Sicily, the capture of Palermo, and the subsequent drive along the island's northern coast toward Messina.

5. Samuel Eliot Morison, History of United States Naval Operations in World War II, vol. IX: Sicily–Salerno–Anzio (Boston: Little, Brown, 1954), 42–47. For coastal operations, Ranger landings along the north coast near San Fratello, naval gunfire support, and Axis air attacks on shipping.

6. Wesley Frank Craven and James L. Cate, The Army Air Forces in World War II, vol. II: Europe: TORCH to Pointblank (Chicago: University of Chicago Press, 1949), 248–250. For fighter operations supporting amphibious landings and air cover along Sicily's northern coast.

7. Ronald Lewin, Montgomery as Military Commander (London: Batsford, 1971), 209–212. For Eighth Army operations on the eastern side of Sicily and the friction between British and American axes of advance.

8. Omar N. Bradley, A Soldier's Story (New York: Henry Holt, 1951). For the broader Allied operational context in Sicily and commentary on coordination challenges between ground and air forces.

9. Carlo D'Este, Patton: A Genius for War, 562. For the final phase of the race to Messina and Seventh Army's arrival ahead of Eighth Army.

10. Wesley Frank Craven and James L. Cate, The Army Air Forces in World War II, vol. II, 248–250; Samuel Eliot Morison, Naval Operations, vol. IX, 42–47. For the 11 July 1943 friendly-fire incident involving Allied airborne transports and the subsequent tightening of air-ground coordination procedures.

CHAPTER 14

CELEBRATING IN SICILY

The Sicily campaign was barely finished when the war turned strange.

On the coastal strips near Licata, tents went up not for a bivouac, but for a show. Trucks rolled in carrying lights and sound gear instead of ammunition. Out stepped Bob Hope, Frances Langford, and Tony Romano, a touring troupe from another world.

Figure 14-1. From left: Hal Block, Bob Hope, Barney Dean, Lt. Gen. George S. Patton, Frances Langford, and Tony Romano. Sicily, August 21, 1943. Public domain.

Men who had not seen a woman outside a pin-up in months jammed shoulder to shoulder. Frances sang torch ballads. Colonna mugged and Bob Hope cracked lines about the dust, the chow, and the brass. Laughter rolled across ground hardened by weeks of sorties and beach assaults.

Even General Patton appeared, striding to the front row where photographers aimed their lenses. The timing was deliberate. In the previous couple of weeks, he had struck two hospitalized GIs in a burst of rage, incidents that were already moving up the chain to Eisenhower. Now, with Bob Hope at his side, Patton shook hands, grinned, and allowed himself to be seen. Hollywood sandpapered the edges of his legend, and Patton understood exactly what that meant.[4]

Algiers

For Watts and the 59th, the USO show was more than a diversion. It became a mission. On 24 August 1943, his log carried an unusual entry. "Escort, Hope-Langford flight: Licata to Algiers."[1]

After the USO show, the troupe's C-47 lifted off the same airstrips that had launched convoys into Messina weeks before. This time the cargo was laughter and glamour. Watts bracketed it with his section like any other bomber escort. Two P-40s above and two sliding in off the wingtips. Brownings still armed, radios still crisp, the flight felt routine and bizarre at the same time.

Below, the Sicilian beaches were still littered with wrecks of landing craft and tanks. Above, a transport carried America's entertainers knowing they were safe because pilots like BK flew alongside in serious fighter planes. They droned west across the Mediterranean, delivering the troupe back to Algiers as if returning them to another world.[2]

Algiers was not only a rear area and a stage for celebrity. It was the headquarters of Twelfth Air Force, where the paper war eventually caught up with the flying one. On 26 August 1943, BK received the Distinguished Flying Cross for extraordinary achievement as a P-40 pilot in the North African theater. The citation noted hazardous bombing missions, sea sweeps, reconnaissance, strafing, and escort sorties carried out from Morocco through Tunisia and Sicily.[3]

On paper, it was another line and another ribbon. In reality, it marked that the long run of low-level work had been seen and counted.

A few days earlier, Hope and Langford had stepped into waiting cars amid applause. Watts and his crew had unbuckled in silence and walked back to their tents and cots. In the same month he had strafed Milazzo and downed enemy fighters over Sicily, he had also escorted comedians and singers under armed cover. The absurdity of it all lingered.

Forward to Paestum

Recognition did not mean rest.

By 10 September 1943, Watts was moving forward again. Bone to Cape Bon. Cape Bon to Termini Imerese. Orders stacked faster than souvenirs.[1] The next beaches would not be Licata or Gela. They would be at Salerno, on the Italian mainland, where Lieutenant General Mark W. Clark's Fifth Army would attempt a landing that hung by a thread.

The interlude closed as quickly as it had opened. The war shed its costume and went back to work.

Notes

1. Blanchard K. Watts, Pilot Flight Logbook (Aug–Sep 1943, author's collection), entries for 24 August and early September routings.

2. U.S. Army Special Services, "Itinerary, Bob Hope USO Tour 1," HQ AFHQ, 19–31 Aug 1943, RG 160, National Archives II; National Archives Still Pictures, series 111-SC (Hope and Langford performances at Licata/Gela, August 1943).

3. Distinguished Flying Cross citation, Headquarters Twelfth Air Force, 26 August 1943; Blanchard K. Watts decorations file (family collection).

4. Carlo D'Este, Patton: A Genius for War (New York: HarperCollins, 1995), 556–61.

5. George F. Howe, Northwest Africa: Seizing the Initiative in the West (Washington: CMH, 1957), 194–200.

CHAPTER 15

AVALANCHE

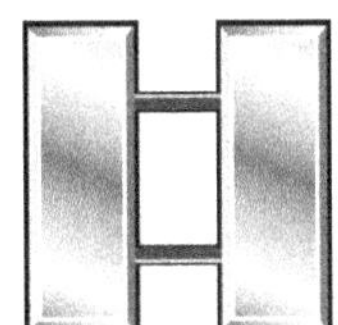

The sea off Salerno was crowded with hulls and smoke in mid-September 1943. Allied convoys filled the horizon, men and machines bound for the Italian mainland. Operation AVALANCHE had begun, though few of the men flying cover needed the name. They knew only that the beaches ahead were narrow, the hills were close, and the Germans were waiting.

Mussolini was gone and Italy had surrendered on paper. In the ground above the beaches, the defenders were German, and they intended to drive Lieutenant General Mark Clark's Fifth Army back into the water. British and American corps fought side by side under Clark's command, a coalition tested at the edge of the surf.[1] It might have been Patton's army if not for the storm that followed him in Sicily. Now Clark held the lodgment, and the margin for error was thin.

Captain Blanchard King Watts lifted his P-40 from the strip at Termini Imerese in Sicily with his flight, the routing already penciled onto his wristboard. His logbook remained spare, but the mission set had shifted:

- 13 Sep. Sortie. Salerno landing
- 13 Sep. Sortie, 1:00. Bomb traffic inland, Salerno[2]

Holding the Beachhead

From above, the landings looked fragile. The beaches near Paestum were narrow and exposed, backed by hills that folded inward like a trap. German artillery sat in every crease. On 12 and 13 September, panzer and panzergrenadier units drove almost to the water's edge. Destroyers fired into them like floating batteries into the bluffs. Fighters strafed armor in front of American infantry lines.[3]

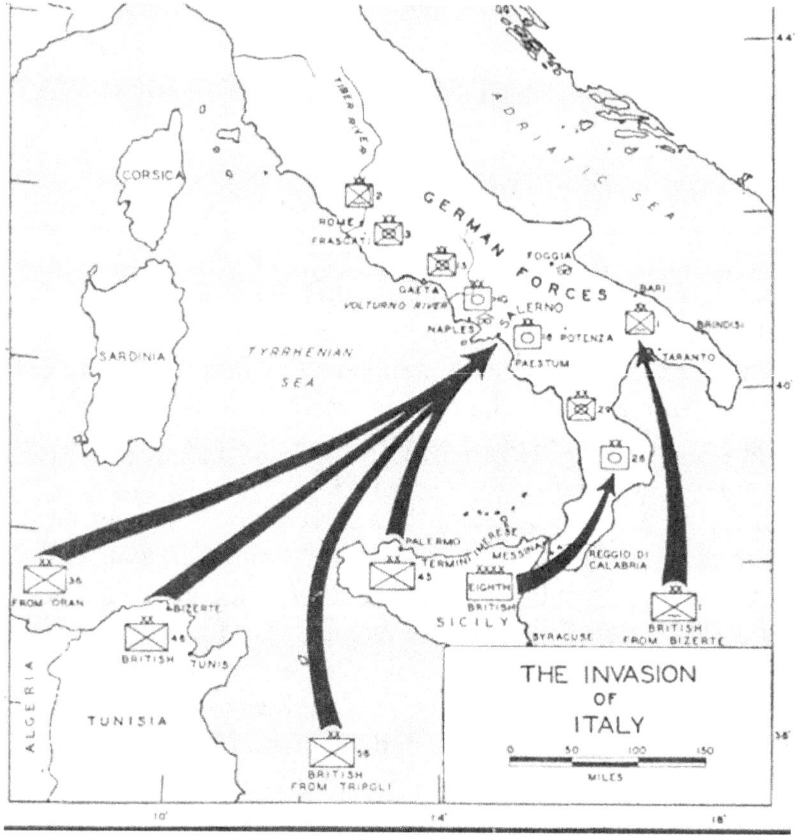

Fig 15-1 U.S. Army. Invasion of Italy, 1943 [map]. U.S. Army Center of Military History. Accessed 10 September 2025. https://commons. wikimedia.org/wiki/File:Invasionofitaly1943.jpg. Public domain.

The 59th flew relentless cycles. Combat air patrols over the beaches, convoy protection offshore, then inland sweeps to choke German traffic pushing toward the lodgment. In Watts's log the phrase was blunt.

"Bomb traffic. North Salerno"[2]

Behind it lay steep dives into flak, roads stitched with tracer fire, and the constant pressure of flying low enough to matter.

Loss came quickly. On 13 September, Watts noted in his log that one of his replacement pilots, 2nd Lt. Charles G. "Franko" Frank of the 59th Fighter Squadron, failed to return. Frank had was an early replacement in North Africa. Later confirmation from the 33rd Fighter Group would list him as killed in action that day.[6] He was one of the small group that went to Agadir in February.

Another Victory

Salerno brought new opponents. The Focke-Wulf 190 appeared low and fast, its radial engine and heavy armament built for ground attack. One came in hard over Allied columns near Paestum, cannon flashes stitching the road.

BK rolled in behind it, closed the distance, and brought all six Brownings to bear. Smoke and flame spilled from the fuselage. The fighter slammed into scrub just beyond a low rise behind the bivouac area.

He wrote it the way he always did:

15 Sep. Patrol   1 Fw 190[2]

In the margin of the log he penciled a small red swastika. It was his third confirmed victory in a month. Not a boast. A recorded event. Each mark another rung climbed, from careful wingman to combat leader.

Expanding the Fight

Once the beachhead held, the tempo did not ease. On 19 September, Watts escorted bombers striking German positions around Salerno. On 23 September, his section swept roads near Ariano to cut traffic north of Naples. Flak rose in sheets. Fighters tightened formation and pressed through.[2]

Even peripheral missions carried weight. On 28 September, BK patrolled the waters off Capri, covering Navy minesweepers clearing channels into the Bay of Naples. Minesweeping was slow, exposed work. Without fighters overhead, the thin-skinned Allied vessels would have been easy prey.

By month's end, the line had held. Clark's battered forces dug in, then linked with Montgomery's Eighth Army advancing up from Calabria. Together they began the slow push toward Naples.[5]

Watts's log read plainly

- 19 Sep. Escort bombers, Salerno
- 23 Sep. Bomb traffic, Ariano
- 28 Sep. Patrol sweepers, Capri[2]

Strung together with hundreds of others, those entries formed an aerial scaffold that kept AVALANCHE intact until the ground forces could move.

Fig 15-2 Sgt. Bernard Reed, BK's Crew chief since Thelepte, and BK Watts alongside BK's P40 named "Cotton Duster" Fall 1943, Italy. Annotations by BK Watts (Authors Collection)

The Cotton Duster

Late in September, the pace slackened just enough for something human to happen on the flight line.

Earlier that month, he had cycled back through Pantelleria to trade tired airframes for newer P-40Ls. Back on the ground near Paestum, his crew chief, Bernie Reed, found an opening.

While the captain was buried in briefings and paperwork, Reed scrounged paint and a brush from places that did not ask questions. In the quiet hour before evening, with the armorers finished and the line settled, he climbed up near the nose and lettered two words on the port cowling in blunt white capitals.

COTTON DUSTER.

No cartoon. No teeth. Just the name Watts had been using all along.

When BK came out to dispersal, he stopped short.

"Sahrnt Reed," he said in his North Carolina drawl, half accusation, half amusement, "what the hell'd you do to my plane?"

Reed was already there, one boot on the tire, elbow on the wing root like a man leaning on a fence.

"Just brought her back to herself, sir," he said. "Figured it was time she knew her own name."

He studied the letters, the faint brush marks, the slight hesitation on the second T. Someone passed him his camera. Reed didn't pose so much as settle into place, still half on the airplane, while Watts leaned back against the warm cowling. No one said much. The engine ticked as it cooled. The moment closed on its own.

It wasn't ceremony. It was recognition. For the first time, the airplane, the pilot, and the war all seemed aligned under a single name.

The work would continue. Harder ground lay ahead.

Notes

1. Samuel Eliot Morison, History of United States Naval Operations in World War II, vol. IX: Sicily–Salerno–Anzio (Boston: Little, Brown, 1954), 68–72.

2. Blanchard K. Watts, Pilot Flight Logbook (September 1943, author's collection).

3. George F. Howe, Northwest Africa: Seizing the Initiative in the West (Washington: Center of Military History, 1957), 190–93.

4. Carlo D'Este, Fatal Decision: Anzio and the Battle for Italy (New: HarperCollins, 1991), 36–39.

5. Ronald Lewin, Montgomery as Military Commander (London: Batsford, 1971), 220.

6. 33rd Fighter Group Combat Digest, reunion edition, compiled and edited by William Usher (SSgt., 33rd Fighter Group), Gore, Oklahoma, 1984. Confirms the loss of 2nd Lt. Charles G. Frank, 59th Fighter Squadron, killed in action on 13 September 1943.

CHAPTER 16

FROM NAPLES TO NEW YEAR'S

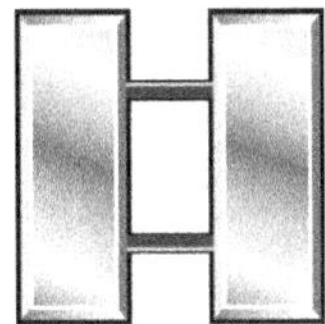

Naples announced itself with smoke and silence. On 2 October 1943, Captain BK Watts crossed the smudged crescent of its bay, flying combat air patrol over cranes that looked like skeletal ribs against a charcoal sky. Half-sunk freighters lay canted at their moorings where German demolition teams had scuttled them. Oil fires clung low over the water like weather. It was a prize seized half broken. The Allies finally owned a deep-water port that could inhale Liberty ships by the dozen and exhale an army up the rocky coast of Italy.[1]

Three days later, on 5 October 1943, Watts was back in the cockpit, flying two bombing runs over the northern reaches of the Volturno River.[3] The Allied advance pressed hard against German defenses anchored on rivers and ridgelines. These sorties targeted supply routes and positions around the Volturno crossings and the town of Teano. Each pass over that river was a calculated risk. Pilots dove through flak to deliver ordnance meant to soften resistance and clear the way for infantry clawing north.

100

On 6 October a milestone arrived disguised as routine. On a strip near Paestum, dust lifting in propwash, Watts lifted with his section to shepherd medium bombers off the field. "6 Oct. Escort, 2:00. B-25s out of Paestum."[3]

109

59th Squadron, 33rd Group, Paestum, Italy
(Salerno Beachhead)
October, 1943

Sitting, L to R

1. Grey L. Neely
2. Ibrie M. Beatty
3. Berry J. Duncan
4. Russell H. Hooker (Doc)
5. Donald A. Halliday (Intel)

6. Charles H. Duncan
7. Blanchard K. Watts
8. Minar M. Dervage
9. D. A. Jackson
10. James L. Reed

Kneeling, L to R

1. John W. Goodrich (Exec)
2. Frederick Zeidler
3. Bernard A. Byrne
4. Harold A. Seestadt (Maint)
5. John C. Makely

6. Robert A. Williams
7. Morgan S. Tyler
8. Ralph L. Antonides
9. GAIL BALLARD
10. Irving P. Tunis

Standing, L to R

1. James P. Finley (Intel)
2. Jack A. Blacker
3. Robert C. Campbell
4. John V. Bach
5. Yonkers (Arm)
6. Harry A. Moyer

7. Alex E. Ristich
8. Glen E. Stewart
9. BILL PARKER
10. Graham E. Fogg
11. Edward K. Dennehy

Fig 16-1 & 16-2 59th Fighter Squadron Officers with legend. Annotations by BK Watts. (Author's Collection)

It was his one-hundredth combat sortie with the group. Underlined for effect. Written in the hurried hand of a pilot who had flown from TORCH through Salerno and now crossed the century mark with the same restraint he brought to formation turns. One hundred times airborne under enemy sky. One hundred landings back into dust and mud. Quietly, the cadet from Hicks Field had become the veteran who survived, and thrived, in three theaters.

Into Naples

The approach to the city was a geography of damage. Villages amputated. Bridges blown. Campaniles toppled. Olive groves seeded with mines. Fifth Army pressed north along the coastal plain while Montgomery's Eighth Army leaned up the Adriatic. Over it all, Field Marshal Kesselring conducted a measured retreat, building defenses by phase rather than panic, charging the Allies in time and blood for every mile of road they took.[5]

From the cockpit, Naples mixed ruin and promise. Shattered piers jutted like fallen colonnades. Smoke-glossed water reflected battered domes. Yet, cranes could be rewired and quays could be cleared. The harbor was big enough to shorten the continent's supply line. BK's log, as always, held to the essentials.

The River of Obstinacy

Twenty miles north, the land buckled into a problem with a name. The Volturno River. Broad and coiling, with high banks ideal for interlocking German flak. Behind it, the ridgelines stepped toward Cassino.

To the infantry it meant black-water crossings under machine-gun fire.

To the artillery it meant duels in limestone valleys.

To the Twelfth Air Force and the 59th it meant an unglamorous mandate. Cut everything the Germans could drive. Break every span they could cross and keep a fighter lattice over Clark's troops.[4]

The squadron called it, grimly and admiringly, "the obstinacy." The word fit both the river and their enemy.

BK's October log shows the shift from beach cover to road kill-

ing, each entry a hammer strike:
- 10 Oct. Recce/Strf, 1:50. Roads hit, heavy flak. Aborted to weather, retasked next day.[3]
- 18 Oct. Bomb traffic and troop concentrations north of the Volturno at Vairano.[3]
- 20 Oct. Bomb road and rail at Vairano.[3]
- 26 Oct. Bomb gun positions at Vairano.[3]

Vairano Scalo and the upper Volturno choke points became familiar as a gunsight. The rail cut gleamed like a wet thread between hedgerows. Culverts flashed white when bombs were dropped. Trucks corkscrewed smoke around cypress trunks. Flak appeared from the same olive patches day after day. Invisible until it was not.

On the deck, the P-40 became a hammer. A reliable tool. Briefings had reduced geography to landmarks. Turn at the stone bridge. Watch the orchards. Eighty-eights buried there. Egress low and west. Do not silhouette on the ridge. It was discussed with carpenter's practicality, as if they were framing a roof, not strafing a column.

Clipboard, Handshake, Command

Eleven days after his hundredth sortie, a jeep bucked through puddles and stopped beside operations. A clipboard changed hands. The outgoing commander had orders in his pocket.

On 17 October 1943, the sentence landed without ceremony.

"The squadron is yours, Captain."[2]

Just like that, BK Watts became commanding officer of the 59th Fighter Squadron.

Command added weight to everything. Crew chiefs straightened when he walked the line. Lieutenants watched his eyes as much as the map. The operations clerk brought rosters and weather. Intelligence brought photographs and the faces of men after hard landings. Debriefs meant absorbing blame that was not always his. Writing letters home to grieving widows, however, required a steadier hand than the control column ever did.

North to San Pietro

The Volturno crossings held. Fifth Army leaned into the Mignano Gap. New city names appeared on the board. San Pietro Infine. Cassino. Places that would grind divisions into dust.

The air tasking widened south of Cassino. Bridges and switchbacks repeated like stanzas.

"5 Nov. Bomb traffic south of Cassino."[3]

The Luftwaffe thinned by daylight. The flak did not. It climbed in invisible staircases. Linger too long and the earth rose to meet you. Cannon shells tore ragged bites from wings that ground crews patched into quiltwork. A pilot might return holding a glove blackened by cordite and say, evenly, "Came right through the vents."

Winter's First Grip

November gave way to storms. Strips became mud heaps. Boots sank calf deep. Tents sweated. Instruments fogged on climbout and iced on letdown. The 59th still flew, but incrementally. Weather vetoed more sorties than the enemy and their routine settled in.

Stoves were built from spent shell casings. Mess lines produced tea and something that was meant to be stew. At dawn, engines started under low cloud, the sound muted, as if they were wrapped in canvas.

Reed's maintenance grew meticulous. Spark plugs fouled constantly. Magnetos sputtered. Propeller blades filmed over with ice that crept along the leading edge and stole lift by inches.

December arrived not with drama but with a long exhale. Kesselring husbanded fuel. Luftwaffe daylight appearances were few and far between. Missions still posted although many escorts met no opposition.

By December the war shifted into a different register.

On 10 December, Paestum was weathered in. The strip vanished into low cloud and rain, engines coughing themselves awake only to be shut down again. Nothing moved except men pacing the flightline and ground crews staring at the sky.[2]

On 13 December, Watts logged a "routine test flight." The phrase hid the truth. He lifted into a ceiling that pressed low and damp against the canopy, instruments fogging as fast as they could be wiped clear. The airframe felt heavy, sluggish, every response delayed by cold oil and thicker air. He stayed close, circled once, twice, listening for a note that didn't belong, then brought the P-40 back in with the caution reserved for machines that might quit without warning.[3]

By 30 December, Paestum was still marginal. Watts flew three test hops, short and deliberate. Up, around the pattern, back down. Again. And again. Slow time, he wrote. Not boredom, but vigilance. Each flight was a question asked of the airplane. Each landing was an answer he wanted to hear before he trusted it with ordnance and men again.[3]

12 Clusters

By winter, the paper war caught up to BK again.

Citations did not arrive in sequence with the danger that earned them. They lagged weeks, sometimes months, moving through headquarters long after the engines had cooled and the mud had dried on their boots.

In February 1943, during the North African campaign, Gen Doolittle had presented Watts with his Air Medal for meritorious achievement in aerial combat. That ribbon had already traveled with him through Tunisia and Sicily, its meaning having been absorbed into routine rather than ceremony.[3]

On 19 November 1943, as the 33rd shifted between Paestum, Naples, and Cercola, orders arrived awarding him his twelfth Oak Leaf Cluster to the Air Medal.[3] It marked no new chapter in the war. It closed accounts already paid with heavy risk.

From Naples to Vairano to the shadow of Cassino, Watts's war had widened. What began in seconds of tracer fire had stretched into months of planning. What began as flying the mission had now become ownership of the consequences.

A new year approached. The work did not pause.

Notes

1. Samuel Eliot Morison, History of United States Naval Op-

erations in WWII, Vol. IX: Sicily–Salerno–Anzio (Boston: Little, Brown, 1954), 90–94.

2. Blanchard King Watts, War Diary (1 Jan 1944 entry, author's collection).

3. Blanchard King Watts, Pilot Flight Logbook & Squadron Orders (Oct–Dec 1943, author's collection), including: 6 Oct (B-25 escort from Paestum; 100th sortie); 10.18, 10/20, 10/26 Vairano strikes; 11/5 bomb traffic south of Cassino; 17 Oct squadron C.O. designation; Special Orders awarding Air Medal, 20 Feb 1943, and Twelfth Oak Leaf Cluster to the Air Medal, 19 Nov 1943. (author's collection)

4. Carlo D'Este, Fatal Decision: Anzio and the Battle for Italy (New York: HarperCollins, 1991), 55–61;

5. Derived from the combined accounts in Notes 1–4 above

CHAPTER 17

VILLAS, VOLCANOES, AND CASSINO SKIES

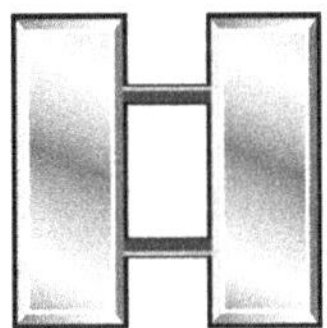

By January 1944, the war had begun to leave room for accounting.

Not the kind done on maps or in operations tents, but the quieter kind that happens when a man survives long enough to look back. Notebooks were scarce in Italy, paper rationed and reused until it fell apart. What BK found instead was an A4 sized accountant's ledger, maroon-covered, green-taped at the spine, ruled for debits and credits rather than memory.

Inside the front cover, he wrote the provenance carefully, as if ownership mattered:

"Bought in Naples, Italy, by Ibris M. Beatty for a total of 200 lire. To be used for notes, experiences, diary and such along that line (sic) of Blanchard K. Watts At this time, New Year's of '44, stationed at Paestum, Italy in the 33rd Fighter Group - flying P40's - helping push the German Army back toward Rome..."

Then, without flourish, he recorded the moment.

Diary: 1 January 1944

"I'm rather surprised I saw this new year at all.

Flew a total of 107 combat missions over Tunisia, Pantelleria, Sicily, and Italy. Got three confirmed victories, dropped some great number of bombs, and shot up some trucks, armored vehicles, gun positions, camels, and jackasses, all with no injury to myself.

Advanced from wing pilot to squadron commander, and saw a

lot of my friends go in different ways. Many returned to the States. Arterburn, Brown, Marks, a few others and I are the only pilots left in the group that came over a little over a year ago. Brown is group operations officer, and Marks is in my squadron.

We expect to return to the States sometime in the near future.

Went to bed early last night. No New Year celebration for me. Had mine over the latter part of last month. Anti-aircraft guns seemed to have blasted away all night, so I just as well stayed up for the sleep I got.

Most of this afternoon making necessary preparations for the move to Cercola tomorrow."

That was the first time BK sat down to record his wartime experience beyond the entries in his flight log.

Villas, Rain, and Lava

Three days later, the 33rd FG commandeered three villas in Cercola, at the foot of Vesuvius. A coin toss settled the housing debate. The 60th drew the cramped residence. The 58th and 59th shared the largest one and the other villa became a club and mess. Men who had lived in tents since Tunisia now argued over rugs and scavenged draperies from the street markets in Naples. BK recorded the novelty with dry surprise: *"First time any of us have lived other than in tents since entering combat. Much interest in redecorating."*2

The comfort that was afforded by nice things, however, felt thin. On 6 January, a fever pinned Watts to a cot. Rain drummed the tile roof while Vesuvius brightened the night, lava sliding downslope, sparks lifting and dying in the downpour. From his window he watched volcanic fire and listened to thunder work the beams. He did not fly again until the tenth yet the mountain still glowed like a forge at the edge of his vision.

Rooftop Duels and Shattered Knees

While Watts lay sick, the war moved on. On 8 January, Lieutenant Glenn Stewert tangled with an Fw 190 over Cassino's rooftops, twisting among the chimneys and smoke for twenty minutes until both pilots ran out of ammunition and broke away alive. No victory. Survival for both seemed like a worthy compromise.

That same day ended harder. Lieutenant Howard Young took a direct flak hit. He completed one final bomb run before bailing out over friendly fields, alive but with his right knee destroyed. His war was finished. Relief and grief bent together. Rivalry among fighter pilots ran deep, but the empty cot that night needed no explanation.

Villas, Helmets, and Dogs

By 12 January the 59th occupied its own villa, its broad stairs and vaulted ceilings echoing every step. Pilots hauled in sofas, traded fuel rations for carpets, and debated the worth of a chipped marble bust pulled from the cellar.

That evening, a crate of British leather flight helmets arrived. Watts ordered them issued to the front-line pilots of the 59th. Group headquarters countermanded him. Angry, he surrendered the crate. Command, he learned again, meant absorbing the slight without losing control.[4]

FIg 17-1 Mascots of the 59th: Pluto and Raunchy outside a squadron tent, Sicily 1943. Annotations by BK Watts (Author's Collection)

Cercola also gained two fixtures: Pluto and Raunchy, stray

dogs adopted in Tunisia. Pluto patrolled the corridors like an inspecting officer. Raunchy tracked kitchens with professional focus. For the men whose friends vanished with each mission report, the dogs' uncomplicated presence steadied the villa more reliably than liquor ever did.

Fire and Furniture

At dawn on 13 January, Watts led the 59th against the ammunition dump at Villa Atina. Bombs touched off a chain of secondary explosions that drove black columns into the sky. Observers said the valley looked like a foundry waking up. Watts wrote five restrained words: *"Large fire. Smoke towered over valley."*5

By afternoon he was back in Cercola negotiating with the villa's owner over carved chairs and wine cabinets. Allied requisition was the final word. They compromised and the furniture and keys stayed with the Americans. Italy often resolved that way, bombing by daylight and moving pianos at dusk.

Discipline and Dances

On 14 January the squadron reached the field late. Colonel Loring F. Stetson Jr., now commanding the 33rd, waited on the tarmac and dressed Watts down in front of his men. The rebuke carried across aluminum skins on the flightline. The mission, however, launched on time. Watts noted it with clipped relief: *"On time, though plenty rushed."*6 That afternoon, he restored the balance by recommending eight lieutenants for promotion.

The next day swung between violence and farce. In the morning haze, the squadron cratered a rail depot and strafed a freight engine until its steam boiled through its wrecked pistons. Three aircraft returned torn by flak. That night, however, there was a formal dance in Naples with too many officers and too few women. The Italian civilians still wary of U.S. uniforms. Watts's verdict: *"Dance poor. More men than women. We remain at war, not at a party."*7

Vesuvius and the A-20s

On 16 January Vesuvius rumbled again as the 58th escorted

A-20 Havocs toward Atina. The Luftwaffe came hard, more than twenty Fw 190s and Me 109s slashing through the formation. Lieutenant Robert Campbell destroyed a 190 cleanly, but an A-20 fell burning away from the box.

January 1944 | The War Beyond the Villas

While Watts wrote in a ledger bought on a Naples street and flew from muddy strips around Paestum and Cercola, the war across Europe was tightening.

Italy: Allied forces were stalled south of Rome along the Winter Line. German defenses at Cassino blocked the Liri Valley, forcing a grinding mountain campaign. Fifth Army prepared a high-risk maneuver to break the stalemate with a landing behind German lines.

Operation SHINGLE: Plans were finalized for an amphibious assault at Anzio in late January, intended to outflank Cassino and force Kesselring to abandon his positions. Air units like the 33rd Fighter Group would be critical in covering the landings and isolating the battlefield.

Eastern Front: The Red Army's winter offensives continued to push west. German reserves that might have reinforced Italy were pulled east instead.

Britain and the United States: Air planners pressed strategic bombing of Germany while still feeding aircraft and pilots into the Mediterranean. The air war had become a balancing act between theaters.

By January 1944, Italy was no longer a sideshow. It was a pressure point. What happened over Cassino and Anzio would decide whether Rome fell quickly or at terrible cost.

Blame for the loss of the A-20 followed immediately. It had been the 58th's rotation. Whispers spread that a fighter had slipped the net. For escort pilots, nothing cut deeper.

The next day a party from the 33rd reached the wreck. The

holes in the A-20 were ragged, far too large for .50-caliber fire. Flak had killed the aircraft. Watts recorded the finding with controlled relief: "Initial blame: fighters. Inspection shows flak. Group record still perfect."[4] Meaning the 59th's record in Italy.

Vindication did not lighten the wreckage. A perfect record for the group still meant someone else's luck had failed.

Belching Earth, Empty Skies

On 18 January Vesuvius threw cinders again while Watts flew two bombing sorties along the Gaeta coast, passing his 110th combat mission. The entry barely rose above routine.

The following day briefers warned of massed fighters north of Rome. The group launched to escort B-26s. They met nothing. No contrails. No flak. Only blue sky and cold. Intelligence later suggested the Luftwaffe was holding strength for a seaborne strike at Anzio. The emptiness was temporary.

Notes

1. Blanchard K. Watts, *War Diary*, 1 January 1944, author's collection.
2. Ibid., 3-5 January 1944.
3. Blanchard K. Watts, *Pilot Flight Logbook* and diary entries, 8 January 1944; see also Carlo D'Este, *Fatal Decision: Anzio and the Battle for Italy* (New York: HarperCollins, 1991), 62–63.
4. Watts, *War Diary*, 12 January 1944.
5. Ibid., 13 January 1944.
6. Ibid., 14 January 1944.
7. Ibid., 15 January 1944.
8. Samuel Eliot Morison, *History of U.S. Naval Operations in WW II, vol. IX: Sicily–Salerno–Anzio* (Boston: Little, Brown, 1954), 121–22; corroborated by 33rd FG *Combat Digest*, entry for 16–17 January 1944.

Chapter 18

The Beachhead and the Exit

A year earlier, on New Year's Day 1943, BK Watts had been a green lieutenant in Tunisia, breathing cordite for the first time. Now, in late January 1944, at twenty-four, he stood on the threshold of becoming Major Watts, commanding a squadron of veterans whose tab in lives had long since outpaced their tally of sorties.

The war diary he kept in an Italian accountant's ledger remained clipped and factual, but behind its restraint lay the weight of command. Each line covered combat, reprimands, dances, hospitals, accidents, and the fragile beats of ordinary life taken between missions.

The Hospital and the Whiskey

For the 59th, the buildup to Anzio was as human as it was operational. On 20 January, Watts stopped first at the 17th General Hospital, where Howard Young lay pinned in plaster, his knee shattered beyond repair. Doctors were certain he would never fly again. The ward smelled of antiseptic and resignation.

Watts left him with jokes and gossip from the villa, but he carried something else back through Naples. Two cases of sortie whiskey. Not indulgence, but insurance. A way to steady men about to fly into something worse than Sicily. Leadership, he believed, meant doing what he could so his people could do their jobs.

That evening, Cercola held a festa. Bells rang, voices carried, and pilots leaned on balcony rails while civilians briefly reclaimed the streets of a city and country still torn open by war, with no sign of an end.[2]

Daybreak over Nettuno

Watts's watch read 04:00 on 22 January when he shook the billets awake. Eight pilots, bleary-eyed and unwashed, crowded the villa hallway. He split them into Red and Blue flights, four ships each, then drove them through moon-lit citrus groves to the line shack.

Frost filmed the wings. Breath hung pale in lamplight. The P40 engines caught one by one, propellers flashing like slow strobes.

At 05:40 both flights lifted off from Cercola, throttles wide, wheels breaking free as dawn softened the eastern horizon. They climbed hard, cleared Vesuvius's shoulder, then bent northwest over a black sea toward Anzio and Nettuno.

Watts held Red Flight high, Blue Flight ten minutes behind. The coast resolved beneath them. Destroyers cut white wakes. Landing craft pushed into surf. Smoke pots dragged gray curtains across the dunes. The orders were simple. Be overhead before the first assault craft touched sand. Keep the sky empty.

They did exactly that. The 59th rode the sunrise without loss. Watts wrote it plainly.

"0815 over beachhead. CAP uneventful. No Jerry. Relieved 0905."[2]

Elsewhere the hour was brutal. The 58th lost 2nd Lt. Russell Schmunk when his engine seized and he ditched into the Tyrrhenian Sea. Boat crews saw the splash but never saw a flare.[3] A pilot from the 315th vanished on a night hop. Another cartwheeled on recovery and left his aircraft broken like a marker on the beach.

The worst came on the strip. Frost slicked the Warhawks during the damp nights. On 23 January, 2nd Lt. Robert Reed, a replacement who had barely arrived, rotated too early. One wing stalled. The P-40 snapped into an uncontrolled roll, struck the ground, and burned. The ground crew reached him seconds too late. The diary recorded it without decoration. "Frozen wings. Fatal."[2]

Another pilot felt his aircraft slide, popped the canopy, and jumped clear as the P-40 skidded into mud and folded its gear. He walked away alive, shaking ice from his gloves.

Lt. Randall of the 60th took flak through a coolant line on coastal patrol, dead-sticked onto a beach north of Gaeta, then waded into the surf to flag down a minesweeper. One more life saved by inches and luck.

By dusk, the winter winds howled off the bay. A final scramble order came and went. Watts wrote the relief cleanly. "2000. Evening mission scrubbed. Great relief to me."[2]

He poured a shot for every man who had pre-flighted before dawn and was still breathing. Outside, artillery flickered south of the villa. Inside, an Italian clock ticked steadily, as if the day had been ordinary.

Frost, Fire, and Fatigue

The days that followed carried both pace and loss. Frost grounded aircraft, then sent others skittering down runways. Lt. Gail Ballard was torn apart by flak on the 26th. His wingman Stewert carried the grief like exposed metal. Missions continued. Bombing by day, strafing supply lines at dusk. Launches as early as 02:30 in fog and ice.

The air war ground on heavily over Italy. From the Gustav Line to Anzio, fighters were, at the same time, eyes, shield, and hammer. Reconnaissance. Escort. Intercept. Control of the air was not optional. It was the condition that kept the ground war alive.

Life intruded anyway. Watts hired an Italian boy to sweep floors and set flowers in jars. The squadron doctor arrived one night with three Italian women and laughter briefly displaced duty. On 25 January, British Major Coffey from a P-47 group observed quietly, feeding rumors of a move north. He vanished from the diary as abruptly as he appeared.

Briefings blurred into dances. Accidents rolled away into gossip. Grief was briefly abated with rumors of rotation home. Above it all, Mt. Vesuvius was getting ready for its own spectacle.

Into Jerry's Midst

On 27 January, the 58th tore into the Luftwaffe "in the midst of Jerry," claiming seven victories. At dawn the next day, the Tuskegee Airmen of the 99th flew with them and added seven more. Over Anzio, the sky layered itself. Ships below. Infantry ashore. Fighters turning and firing above it all.[2][3]

Watts flew, but he also trained. He used the C-78 Bobcat to push his men toward twin-engine habits. Whatever came next would demand more than a single-engine mind.

The future sharpened quickly. Orders posted a group move by 9 February. Pilots penciled "England?" and then imagined pubs and grass fields. On 2 February, after Watts logged a bombing run and a C-78 escort, the truth arrived. The 33rd Fighter Group was ordered to the Far East. Ideas of England dissolved. China-Burma-India replaced them.[2]

Some rotations home finally came. Stewart, Beatty, Smith, Tunis, and Brown held orders that meant survival. The villa split between celebration and a careful, almost guilty silence. Bags were packed in one room while another stayed dark.

Watts carried the heavier line. His promotion to Major was effective 28 January. Not home, but into group staff duties. Not rest, but responsibility. He would shortly help move a fighter group across the world, eastward toward Karachi, one airfield at a time. The aircraft carried everything the group needed: pilots in the cockpits, spares and tools lashed behind seats, records folded into map cases, and the quiet personal cargo that every man pared down to only the essentials. There was no rail transport waiting and no ships to board. What did not fit aboard the C-47's simply did not go. Distance, weather, fatigue, and load limits governed each leg. The war would not loosen its grip just because the map changed.

Ciao Italia

On 5 February 1944, the Naples Officers' Club filled with American voices. Bottles passed and toasts tangled arms together in joyful farewell. Verses for the dead mixed with songs for the living. They were not leaving victorious or beaten. They were simply leaving.

At closing time, someone shouted into the smoke, "Ciao, Italia!"

It landed as a farewell and an elegy together. The ledger turned its page. Vesuvius burned on, waiting for another pilot to watch it glow.

Notes

1. Samuel Eliot Morison, History of United States Naval Operations in World War II, Vol. IX: Sicily–Salerno–Anzio (Boston: Little, Brown, 1954), 120–22.
2. Blanchard King Watts, War Diary (Jan–Feb 1944 entries, author's collection).
3. Blanchard King Watts, Pilot Flight Logbook and AAF Form 5 (Jan–Feb 1944, author's collection).
4. 33rd Fighter Group Combat Digest (1984 reunion volume).
5. Allied Air Forces Liaison Report, "P-47 Tactical Exchange," 25 Jan 1944 (AFHRA microfilm).

PART IV
THE CBI

The China–Burma–India Theater was never a clean front line. It was an accident of geography, stitched together by diplomacy rather than design. From the railheads of Assam and Bengal, across the jungles of Burma, into the Karst mountains of Yunnan, the land was vast, wet, and indifferent. The CBI ranked last in Allied priority, yet the Allied forces there were expected to keep Chiang Kai-shek's armies in the field, hold back a flank of the Japanese empire, and keep China inside the Allied war.

It was less a battlefield than an ordeal. Jungle claimed entire wrecks, swallowing aircraft and crews alike. Pilots fought terrain as often as enemy aircraft. Engineers measured progress in feet cut from mountainsides. Disease, weather, and distance imposed penalties long before combat entered the picture. For Americans arriving from Europe, this was not a continuation of the war they knew, but a different test entirely, one measured in endurance as much as firepower.

For Major Blanchard K. Watts and his squadron, the pivot was abrupt. Weeks earlier they had flown over abbeys and stone towns along the road to Cassino. Now they were halfway around the world, in a theater where the sky itself could be lethal, navigation uncertain, and survival dependent as much on judgment as on weapons.

The accountant's ledger would continue to fill up. The handwriting would stay the same, but the war had changed.

CHAPTER 19

HURRY UP AND WAIT

I spent most of early February packing. Personal kit, odds and
ends, the paperwork of a squadron commander closing the books.
I wrote recommendations, turned over duties, cleared desks, and
watched Capt. Walter L. Moore, Jr. slide into the Squadron Command seat I had warmed. By the evening of the fourth, most of
Naples was in uniform and drunk, tearing a last memory from the
Allied Officers' Club. The 33rd Fighter Group came in loud; the
58th and 59th together owned the place until dawn. We left sandwiches on the kitchen counters and whiskey on our coats.

On 5 February, the 59th was ordered onto transports. C-47s
packed at 10:30 a.m., engines jittering for Tunis. There were never
enough aircraft to go around. The 60th drew the short straw. I
rode into Headquarters' new job, Group Operations Officer. Tried
a hop to Foggia in the C-78 for a new arrival's baggage, turned
back by weather. Naples again that night with staff, the strange
habit of farewell repeating itself.

The days stacked.

6–8 Feb. Each morning: pack, wait, sweat out the transports.
Each afternoon: learn none were coming. Each evening: put away
just enough clothes to walk into Naples. Back by 1:00 a.m. Eat.
Sack. Repeat. I wrote, almost bitter: "First fresh eggs since Africa.
Found them on an LCI." Two transports finally showed for the
58th on the 8th, leaving Group Headquarters staff still stranded.
Half the officers went back into town. I stayed in and felt calmer

for it.

Col. Stetson and Col. Christman managed to get clear on the 9th. By the 11th the only ones who remained were Davidson, Gamble, Brooks, Yost, Chinnock, Ziegler, Anderson, Mautte, and myself. Rain boxed us up. We joked about formal meetings, "Fathers and Daughters," "Sewing Circle," as if a routine could make the time tolerable.

Fig 19-1 BK Watts leaning on the wing of a C-47 preparing to depart for India with unidentified members of 59th FS. Annotations by BK Watts. (Author's collection)

Tunis

On the 12th, three transports struggled through to North Africa. One clipped its wingtip taxiing; Brooks and I sweated it, unsure if the repair would finish before another aircraft could be spared. Two planes made Tunis. We followed into sleet, dumped into cold stone buildings, no heat, bedrolls the only way to fight the cold.

By luck, I found Col. Christman the next day. He knew a little more about what was going on. Tunis seemed cleaner than I remembered, almost reclaimed, rebuilt after the war's churn.

Then tragedy came anyway.

Pluto was run over by a truck. He had been an unusually sharp dog. He will be missed.

On the 14th, we went into Tunis proper. A French sergeant escorted us through the city, into rooms that officers were not supposed to share. We spent the night at the Astoria, a French officers' reserve. We managed it anyway. Hot water that steamed like luxury.

On the 15th, word came that a wedge of us might finally leave tomorrow. My baggage was overweight, embarrassingly so. I piled it on a dolly anyway.

16 Feb. Took off 7:25 a.m., landed in Benghazi 12:40 p.m. We followed the coast the whole way, nothing but bare land rolling back from the water. The gyro failed; we dared not push to Cairo after dark. We bedded down at the airfield, went into town, found mostly ruins and ash. Benghazi was a ghost town, the Italians fled before the British pushed through with Arabs and locals filling in what was left. I slept restlessly.

17 Feb. Did not get off until 11:30 a.m. We waited for a spare part flown in from Cairo to fix the gyro. An hour out we hit a dust storm that swallowed the world. Full instruments, rough air, an hour of it, hands tight on the straps, stomach wrong in the throat. Cairo and Alexandria both reported closed. We turned back, searched in vain for an auxiliary field, found Tobruk instead. El Adem airdrome. British-run, desolate as the moon. A runway, a shack, desert in every direction. Miserable food, a worse night.

18 Feb. Forecast poor. The pilot decided to try for Cairo. Calm air for an hour, then turbulence and dust again. This time not enough to turn us back. Within three days of trying, we made Egypt. Landed at Payne Field. First hamburger in sixteen months. Coke tomorrow, maybe. Beds with sheets. Hot showers. Civilization hit like a slap. Christman and Brooks joined us by nightfall.

We watched Higher and Higher in the movie house and reentered the war by pretending we could laugh.

19 Feb. Rolled out 5:30 a.m., airborne 7:00 a.m. Over the Suez, I glimpsed Bethlehem, Jerusalem, Jericho, the Dead Sea, then the long pipeline of Iraq. Landed H-3 around noon, ate sandwiches. Air again, crossed to Abadan on the Persian Gulf by 3:00 p.m. A good post. PX with cold beer and ice cream, nearly America in miniature.

Karachi

20 Feb. One thousand three hundred miles. Stopped at Sharjah for fuel. Karachi at 4:30 p.m. The end of the thread. We collapsed into quarters, medics probing for every disease in their manuals. New aircraft awaited, rumor had it. P-47s, said with hope. Local boys carried water and washed laundry. Officers dressed out in bush jackets and stitched insignia. We had become caricatures of expeditionary soldiers overnight.

Karachi was a city. Sacred cows took priority over trucks, dignified even in traffic. Shops promised to make anything if you left tomorrow as delivery date. Boots for me, finally. Ice cream, hamburgers, club dinners. Parties rolled nightly, furniture smashed, fists thrown at provosts, the 33rd keeping its name alive, Naples to India.

The trip ends here, with the month finished. On two-hour notice. New American base, polished, clean, engines spun in the hangars. Ten days at the most, I guessed.

The log doesn't include whether I was right.

Notes

- Blanchard King Watts, War Diary, 3–29 Feb 1944 (Author's collection).
- Blanchard K. Watts, Pilot Flight Logbook (auxiliary reference, Jan–Feb 1944 hops, Author's collection)

Chapter 20

Over The Hump

On 1 March 1944, Colonel Loring F."Stets" Stetson, Major Robert Halliday, and Major Blanchard K. Watts boarded a C-54 Skymaster out of Karachi under "Priority-2" orders. They were to scout ahead to Kunming, meet General Chennault's staff, and return with instructions for ferrying the squadron east.[1]

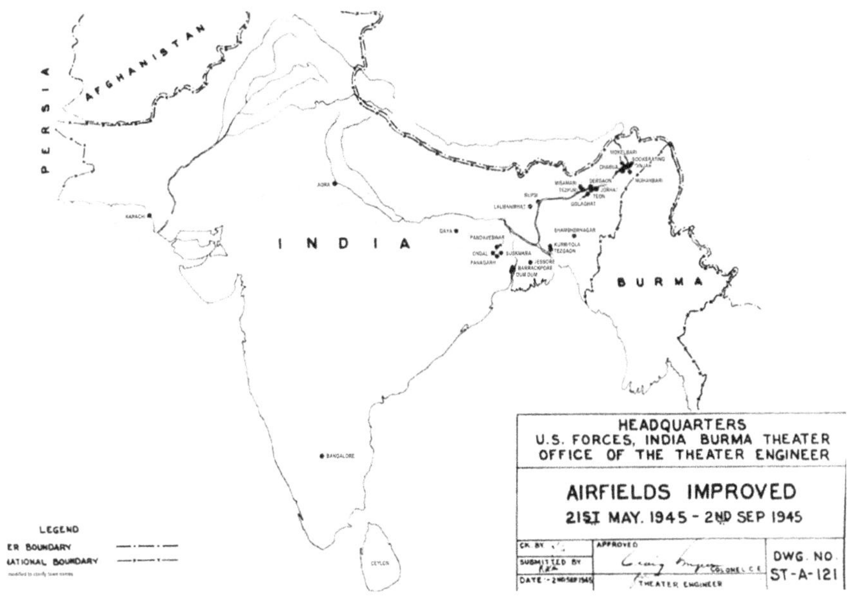

FIg 20-1 Airfields Improved (India 1944-45) Public Domain (Annotations in Orange by Author) https://cbi-theater.com/maps/_Map_Main.html Accessed 11 OCT 2025

That night they stopped at Gaya. The Red Cross mess served hot food, and the men, already weary, stretched out on straw mattresses under a thatched roof. Watts wrote later that he was grateful for spring. The summer heat would have turned the huts into ovens.[1]

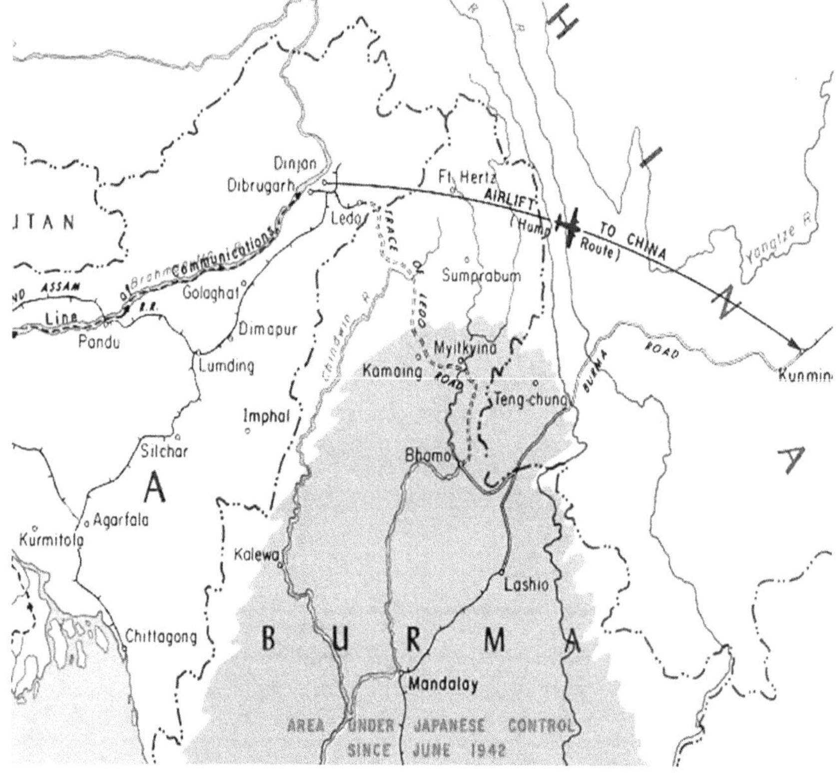

Fig 20-2 MAp: Lines of Communication "Airlift into China" aka The Hump Route (India 1944-45) (https://cbi-theater.com/maps/_Map_Main.html Accessed 11 OCT 2025. Public Domain

Birthday Over the Himalayas

At midnight on 3 March, somewhere above twenty thousand feet, BK quietly turned twenty-five years old. Lightning sketched across the Himalayas. The Skymaster clawed at thin air, and the mountains rose like a wall against the stars. It was his birthday present: survival over the Hump.[1]

They landed in Kunming at two a.m. Hungry and groggy, they found food, then collapsed into their cots for a couple hours of shallow sleep.

At dawn, Watts's birthday continued in a different register. Colonel Stetson roused them at 0830 for check-in at Fourteenth Air Force headquarters. Stetson went on to meet General Chennault, leaving the two majors to their own devices. Watts spent the morning with the 51st Fighter Group, listening to veterans describe the Japanese in the air: sudden vertical climbs, sharp rolls, never turn with a Zero. It was an immersion course in enemy tactics.[1]

Outside the briefing rooms, the lineage was visible everywhere. The toothy shark mouths snarled from engine cowlings, their paint scuffed by dust, hours and sun. The men still wore the habits of Chennault's old Flying Tigers, now folded into the Fourteenth AF but they spoke their own grammar of air combat. There was an ease among them that came from having learned the enemy the hard way. They spoke in shortcuts and warnings, as if the sky itself had already taught the lessons and language was only there to keep men alive.

Afterward, Watts flagged a rickshaw into town. The wiry runner pulled him through Kunming's dirt streets. He felt guilty until he learned he had been charged double the local rate. His diary fumed at the swindle: foreign, dirty, unfair. It was his first lesson in Kunming's economy. The evening was not a total loss, however. For his twenty-fifth birthday dinner, a gift to himself, he found an improbably familiar steak and potatoes meal at a hotel run by the wife of an American doctor.

On the rickshaw ride back, he stared at the airfield itself amid acres of burial mounds. Chinese families were shouldering their dead from the city to the fields. Some of the graves were barely covered with dirt. Others lay wide open. The runway cut straight through a necropolis. It was an eerie reminder of why the Americans were in China.

"We Just Got Here"

On the morning of the eighth, the anticlimax came fast. Chennault was away from headquarters, but his staff delivered the message bluntly: return to India and get the group back here fast.

"Wait... we just got here," Halliday muttered to Stetson.

The reasoning was unyielding. They had proven the Hump could be flown. The jitters were gone. No further reconnaissance was needed. Now it was time to bring their P-40 squadrons east.

For Chennault, the decision was not academic. His fighters were worn thin, losses replaced slowly if at all. Every serviceable airplane mattered, every pilot counted. Japanese pressure along the approaches to Kunming was constant, probing for weakness. The air defense of southwest China ran on margins that would have been unacceptable anywhere else.

After sending wires back to the squadron in Karachi, they left Kunming that afternoon. The message was plain. Move now. The airplanes were needed in the air, not waiting on perfect conditions that would never come.

The three men retraced their steps west. At Agra, Stetson split off for Delhi to handle high-level discussions on the redeployment of the group. Halliday and Watts continued on to Karachi carrying orders that had already been sent ahead.

When they finally arrived back in Karachi, road-weary and dusty from the route, the shift was already visible. The 59th was scrambling to ready its P-40s, not as an isolated unit but as the leading edge of the 33rd moving east. The decision had already passed into action.

Crates stood open on the hardstand. Ground crews worked through spare parts and tools laid out on tarps. Armorers and mechanics moved from aircraft to aircraft, checking fittings, changing tires, adjusting controls, preparing machines that would have no margin once they left the coast. Pilots walked checklists aloud.

Karachi: High-Altitude Trials

Within two days, the fighter line in Karachi grew restless. On 12 and 13 March, P-40s climbed into the desert sky, engines straining toward twenty thousand feet. Oxygen masks frosted. Gauges trembled. Crews scribbled on kneeboards: climb rate, ceiling margin, fuel burn. They had been told what waited above the ridges. Now they carved the numbers themselves. The Hump drifted from myth into something that demanded respect. Get it wrong, and things would go badly.

Given the preparations and Chennault's orders fixed in their minds, they pushed through. At least now they knew how the planes handled altitude, and more importantly how their gear performed: goggles, oxygen masks, clothing. That night, the pressure eased at the Grand Hotel. Steaks, whiskey, and laughter loud enough to drown nerves. The boys drank until two a.m., knowing the ferry east began at dawn.

The Ferry Splinters

On 14 March, twenty-five Warhawks lifted behind a B-25 pathfinder at 10:30 a.m. They were on their way over the Hump. Thirteen hundred miles with a convoy was no small thing. Their first leg overshot immediately, scouring desert wastes for Jodhpur until the strip finally flashed pale below. Two aircraft bogged down in sand. Two never left Karachi. By day's end, Watts's second element had shrunk to nine.

Fig 20-4 Photo 1944 New Delhi: Capt Harry Moyer (R) and Glenwood Eaton (L) both 59th FS, in New Delhi - Annotations by BK Watts (Author's collection)

They refueled quickly in New Delhi, then back into the air for Agra. It already felt rushed, people and materiel shoved forward too fast. Watts sensed trouble ahead.

At Agra, the convoy paused. Watts ran into Staff Sergeant Ben Moore, the wrench-hand from Hicks who used to tap cockpit glass and size up cadets with a smirk, now marooned in India. Moore's crew patched fighters and produced gifts as precious as fuel: T-bones and cold milk. Watts swore it was his first taste of milk in seventeen months. They dined in view of the Taj Mahal, pale marble rising at dusk like a hallucination.

From there, the ferry fractured. Tailwheels collapsed. Engines quit. Storms pinned men at Mohanbari. By 19 March, twenty-three Warhawks straggled together at Chabua. On the twentieth, two flights of eight braved the Hump. Sixteen fighters landed neatly at Kunming. Relief came, followed immediately by dread. Nine were still missing.

Watts began the same routine Major Cochran had used in North Africa after the *Archer* launch. He paced the field, asking pilots what they had heard, what they had seen. Anything on the radio. He decided that if nothing changed by the next day, he would fly back and look for them himself. This was his squadron. It was the least he could do.

Orders intervened. Colonel Amen summoned him north to Yunnanyi, and Watts joined the 25th Fighter Squadron ready room on 27 March. The region felt too remote to warrant raids. The pilots sat talking until one a.m. The war, however, had other plans. At dawn, alarms rang. Props spun. Fighters screamed awake. Then the call came over radios bent with static: "Friendly inbound."

Relief doubled when Watts saw who it was. After landing, one of the 59th, Harry Moyer, buzzed the field and rolled onto the runway, ragged but alive. He had orders for Major Watts, misspelled "Wats." The error drew laughter. At least Harry made it.

Information came in pieces. Three pilots arrived on 28 March. Three more on the twenty-ninth. Only Marks, Deverage, and Schneider remained grounded at Mohanbari.

FIg 20-4 BK Watts' Flying Tigers Patch(L) and 59th FS Squadron Patch (L).
(Author's collection)

Together at Last

By the morning of 1 April 1944, the last aircraft limped in. Engines coughed. Paint was scored raw. The pilots looked hollow from days of weather and fuel stops strung like Christmas lights across the map. They arrived battered, late, but alive. When their wheels touched Chinese soil, the P-40s finally lined up nose to tail across the plateau. For the first time in months, the 59th was whole again.

Deverage arrived later with another flight of three. After getting the planes ready for the next day, the squadron settled into a game of bridge in the last place they had imagined a week earlier.

The following day, Colonel Stetson stepped into full bird rank. The timing felt exact, as if the theater itself had chosen the moment. By dusk, the news moved through tents like sparks in dry bamboo.

That night they slipped into General Gilkeson's quarters to celebrate with Stetson. Maps lay spread on tables. Whiskey bled across Europe and China alike. Men sat on trunks and window-sills. Smoke curled beneath wire-caged bulbs. For hours, laughter rattled shutters. No rain. No alerts. Just men marking survival and reunion.

They had "just arrived" twice before: once onto *Archer*'s deck in 1942, then again in Tunisia. China made them arrive yet again, delayed by weather, paperwork, and distance, dropping into Kunming like dice thrown across a table.

Now, for the first time since crossing the Hump, the Jokers were all in one place. The waiting was over. On the plateau, under paper lanterns and maps ringed with whiskey glasses, their war in China was about to begin.

General Claire Lee Chennault (1893–1958)

Born in rural Louisiana, Claire Lee Chennault entered the U.S. Army during World War I and built his air career in the 1920s and 1930s as a pursuit fighter tactician. At Maxwell Field's Air Corps Tactical School, he became notorious as a fighter partisan, clashing openly with colleagues who insisted that "the bomber will always get through." Chennault rejected the prevailing orthodoxy. In his 1934 manual The Role of Defensive Pursuit, he argued that disciplined fighters, properly directed by observers and early warning, could destroy bomber formations outright.[1]

His views earned him enemies in Washington. Chronic bronchitis, increasing deafness, and constant friction with superiors forced him out of the Army in 1937, retired as a disabled major. He left embittered, convinced that his country had sidelined him for being right too early.[13]

Hired by the Chiangs (1937–41)

Madame Chiang Kai-shek, Western-educated and politically astute, saw in Chennault both an air theorist and a tool for legitimacy. Through her patronage, Chiang Kai-shek hired him as an "advisor" to the Chinese Air Force. With almost no resources, Chennault built an air-warning network out of radios, phones, and even church bells, stretching across rural provinces. He drilled

Chinese pilots in hit-and-run defense, harassing Japanese bombers with every sortie he could scrape together.[34]

The Flying Tigers (1941–42)

By 1940, Chennault's lobbying bore fruit. With Madame Chiang and T.V. Soong persuading Roosevelt's administration, the U.S. quietly allowed him to recruit American fliers from the Navy and Army to serve as "volunteers." The American Volunteer Group (AVG) aka the infamous *Flying Tigers*, flew Curtiss P-40s painted with shark jaws. Their effectiveness from December 1941–July 1942, just as Pearl Harbor plunged America into war, became legend. They downed hundreds of Japanese aircraft with relatively few losses, and their fame in American newspapers made Chennault a household name.[23]

Back In, With Stars

When the AVG disbanded in July 1942, Washington could not ignore reality. Chennault was recommissioned into the U.S. Army Air Forces. This time not as a cast-off major, but as a brigadier general. The AVG became the China Air Task Force (CATF) under his command. In spring 1943, the CATF expanded into the Fourteenth Air Force, with Chennault promoted again to major general.[13]

From an ignored outcast, he had clawed his way back in less than five years propelled by Madame Chiang, by Chiang Kai-shek's trust, and by the AVG's battlefield legend. Washington may have thought him eccentric; but in China, he was indispensable.

Chennault and the 33rd (1944)

By 1944, Chennault's Fourteenth Air Force operated in permanent scarcity. Fuel was rationed, aircraft were worn thin, and spare parts arrived late or not at all. He fought continuously with General Joseph Stilwell over priorities and resources, but his doctrine remained consistent: narrow patrols, close escort, and aggressive interdiction. Within those limits, the Fourteenth Air Force developed a fierce internal coherence.

Into this force came the 33rd Fighter Group. On paper they

were veterans of the European and Mediterranean theaters. Under Chennault, they were absorbed into something older and more symbolic. They became, in effect, Flying Tigers. Watts himself received the leaping tiger patch, wearing it alongside Chennault's men. BK's diary still read "escort, no enemy," but with that patch stitched to his uniform, those routine sorties took on another meaning: Americans over China, guarding the sky.[2][4]

Notes

1. Blanchard K. Watts, *Handwritten Diary* (Feb–Mar 1944), author's collection. Blanchard K. Watts, *AAF Form 5 Flight Records* (March 1944), author's collection.
2. Charles F. Romanus & Riley Sunderland, *Stilwell's Mission to China.* US Army in World War II, China–Burma–India Theater (Washington, D.C.: GPO, 1953).
3. Sam H. McGowan, *The Hump: America's Strategy for Keeping China in World War II* (Annapolis: Naval Institute Press, 2016).
4. Barbara W. Tuchman, *Stilwell and the American Experience in China* (New York: Macmillan, 1970).
5. Claire L. Chennault, *Way of a Fighter* (New York: Putnam, 1949).

Chapter 21

120 More Days in China

By May 1944, time in China had acquired a shape. On paper, the 33rd Fighter Group was on a 180-day loan to Chennault's Fourteenth Air Force. In practice, roughly 120 days remained, and none of them promised resolution. China was not a place where problems were solved. They were endured, absorbed, and carried forward.

Kunming's plateau taught the lesson early. Mud swallowed tires whole. Rain stitched the air to the ground until the sky and the earth seemed fused. The clouds snagged on the hills and refused to move. The main runway had been cut through old burial mounds, the earth giving up its dead so airplanes could pass overhead. Headquarters consisted of bamboo and tin roofing, with borrowed tables and typewriters clattering beneath the steady patience of Chinese interpreters. Supply caravans arrived irregularly, like rumors that sometimes proved true. Each dawn began the same way, with two chalkboards set side by side: weather and war. In China, presence was doctrine for the 33rd. You flew as much to be seen as to strike.[1] [2] [5]

At Pungchachung, the group assembled a workable headquarters from scavenged lumber and flattened fuel drums. Life settled into a pattern no one bothered to call routine. Rice steamed in the mess line. Briefings were held beneath poles that swayed when the wind stiffened. Outside, P-40s and the occasional hulking P-47 crouched under weather that respected neither canvas nor aluminum. The maps on the walls bore thumbtack scars as the front shifted back and forth, sometimes by inches, sometimes by enough to matter. No one spoke of this as temporary anymore. China worked on men slowly, filing them down with persistence.

Fig 21-1 Operations Bldg at Pungchachung 1944 (Author's Collection)

May Nights: Alarms and Cold Sky

The alarms often came without pattern. Sirens shrieked across the plateau during the blue hour, neither night nor morning, with a sound that traveled too far in thin air. One night, with radar reports of a Japanese push toward Ankang, Colonel "Stets" Stetson and Watts lifted into the dark, their P-47Ds climbing until oxygen hissed steadily at twenty-five thousand feet. Masks burned cold against their skin. Ice crusted brows and mustaches. The cockpit narrowed to breath, gauges, and the pressure of waiting.

The plan was simple and familiar. Hold altitude. Wait for attack vectors. Dive if ordered then scatter the formation. Restore the quiet.

No order came. The enemy turned back into the darkness that had produced them.

The pilots descended with emptier tanks and heavier bones. Watts recorded only, "Up high with Stets. No contact. Turned back." In China, half a mission could be spent chasing a ghost, and that readiness itself became a form of attrition.[2] They'd come home tired without having fired a shot, and the fatigue counted all the same.

The Last Joker Standing

On 27 May, a C-47 transport warmed up before the morning mist had burned away. Arterburn slung his duffel. Marks shook hands too long. Hearing. Morgan. Nightingale. Names that had anchored rosters since North Africa, were spoken now with the restraint of men who understood what leaving meant.[2]

Within minutes they were airborne, boots still black with China's clay. They smiled for the younger pilots, saluted their C.O., and looked away when someone said, "Safe travels."

When the dust behind the tailwheel settled, BK understood the arithmetic without counting. He was the last member of the original JOKER Squadron still active in the 33rd Fighter Group.

He walked back to Hanchung's makeshift headquarters, bamboo rafters overhead, maps pinned and re-pinned, coffee still tasting of scorched tin. The sense of continuity tightened around him. From the chalkboard aboard HMS *Archer*, through Morocco, Tunisia, Sicily, Italy and now this emerald plateau, he was the last remaining thread.

He recorded it in his diary, "That leaves me alone of the original pilots that came in on the invasion of Africa in Nov. of 1942..."[2] The brevity carried tremendous weight.

A Narrow Escape

The rhythm of loss continued, indifferent to rank or timing. As Major Halliday rotated back to Hanchung, news of one of BK's pilots dropped like a stack of books onto a table.

Two days earlier, Lieutenant Tyler had suffered a crack-up at Hanchung. The strip has always been short and unforgiving, softened by weather and cut tight against rising ground. On takeoff, something went wrong. Whether it was power, control, or surface, BK never pretended to know. The airplane left the ground poorly, staggered, and struck.

The impact knocked Tucker unconscious. The Warhawk slid into the trees and burned.

Fuel ignited almost at once. Ammunition cooked off in the heat. With no one at the controls, the aircraft became a furnace, with the guns firing themselves as the temperature climbed. Tyler came to inside the cockpit, disoriented, the air already thick with smoke. He clawed at the harness and tumbled free just in time, rolling clear as the fire took full hold.

The ground crew reached him after the ammo was spent. He was shaken, burned, and lucky in ways that did not invite celebration. The airplane was lost. The margin between survival and death had been mere seconds.

It was a reminder written in flame rather than ink. In this theater, the enemy was only part of the danger. The ground itself waited, patient and exacting, ready to take any mistake that flight or weather offered it.

Comedy in Hanchung

Four days later, Hanchung answered the season with an earnest gesture. The city staged a benefit for its airmen, Chinese and American alike. The theater filled with uniforms, suits, and bright local dresses. Familiar Chinese courtesy turned quickly into warmth.

When the curtain rose, a crimson banner stretched across the stage, its translation rendered carefully into English:

COMFORT PARTY FOR PAINFUL ALLIED AIRMEN

The room erupted. Laughter came fast and generously. Their hosts smiled wider, proud that their greeting had landed so forcefully, even if it wasn't precisely as intended.[2]

The evening moved through anthems sung cleanly, Western numbers played on violins and bamboo flutes, and speeches that stepped through interpreters like stones across a stream. Town officials hurried out food until the meal grew to ten courses, ending with peaches, apricots, and cherries. The men ate until they worried, half joking, that the extra weight might keep their aircraft from staggering airborne the next day.

Each guest received a silk scarf and a pastry. In a war that rarely permitted sweetness, the phrase lodged permanently in squadron memory. Years later, at reunions far from China, someone would still toast the "Painful Allied Airmen," and the laughter would come again.

Fig 21-2 Chinese locals and miltary hired to roll out the runway from a grassy field into a sutiable landing strip. Annotations by BK Watts

(Author's Collection)

Into the Wastes: Replacement Flights

Replacement aircraft arrived in June by attrition rather than schedule. P-40Ns came patched up from every stop they along the way. A few aggressive P-47s followed, and then, finally, the first P-51s. They were sleek and unfamiliar, aircraft that seemed to belong to a different phase of the war, faster, heavier, and less forgiving of rough ground.

Watts was tasked with ferrying a detachment east to a staging strip more than two hundred miles beyond Kunming, a raw field isolated from roads, towns, and reliable weather. These flights were not routine deliveries. They were tests of whether the war could be pushed farther outward, whether newer aircraft could be made to live on strips cut from jungle and stone.

The weather turned hostile. Mountains emerged from cloud at impossible angles. Rain erased the horizon. Navigation narrowed to judgment and restraint, to knowing when not to press. Watts later wrote that he barely made it out. Another field that tried to keep him.

The Chinese ground crew met them with lanterns, fuel drums, and steady competence. As they tamped the strip underfoot, patched surfaces, and coaxed engines back to life, the field settled into usefulness. What had been an edge became another rung in the long ladder east.[2]

Bridges and Nerves

From that rung, the reach extended further. Twelve new P-51 Mustangs formed alongside four B-24 Liberators and pointed toward two bridges six hundred and fifty miles away. Even with the Mustang's range, the margins felt personal. No one suggested P-47s for this work. A Jug would have turned back on fumes.

The mission went clean. Weather held. No interception came. Bombs walked across both bridges, methodical and unopposed.

Back at Pungchachung, the bombers landed, refueled, and continued toward Kunming. In the afterglow, a senior general announced he wanted to fly a P-47 forward to inspect the damage. The idea died quietly, smothered by distance, fuel math, and weather. The tension lingered anyway.

China valued symbols. Pilots valued odds.

Fig 21-3 Major Watts near "Travelers Inn" Annotation by BK Watts. Chengtu, China 1944. (Author's Collection)

June 6 in China

On 6 June 1944, while Allied armadas went ashore in Normandy, Watts and the 59th flew escort for Chinese P-40s attacking Yellow River rail crossings. Visibility was excellent. Overcast hung at fourteen thousand feet. Two P-47s rode high above.

The Chinese pilots lingered longer than the Americans preferred. Flak reached once, close enough to matter. The rails were hit. The formation turned north for reconnaissance, then west toward Hsian. Compared to Europe's layered belts of German flak, Watts admitted it felt manageable, even forgiving.

Lunch came at Hsian. New orders followed immediately. The strip lay too close under Japanese reach. The squadron shifted upriver to Paoki, a narrow field perched above the town. The heavy P47 Thunderbolts arrived hard, brake shoes screaming.

Watts wrote only, "It can be done."[2]

That night, orderlies drove a truck onto the airstrip stacked with borrowed beds. The men slept where they fell, under a sky that did not care which continent it hung over.

A radio message the next afternoon ordered them back to Chengtu. Watts understood before the explanation arrived. In Europe, the long-promised invasion had finally begun. News traveled slowly in China, in cables and mimeographed sheets, but the meaning of the invasion of Fortress Europe traveled faster.

Two fronts. One war.

Leadership in Motion

Command continued to move without ceremony. Colonel Christman rotated home, his departure handled with little more than a handshake and a notation on the board. On 13 June, Colonel Stetson followed, slipping out of Kunming almost unnoticed, the man who had carried the 33rd from Italy into China suddenly gone from the morning briefings. There were no speeches. Command in China changed the way weather did, overnight and without sentiment.

That same day, Colonel Whisenand sent for BK. The message was direct. General Gilkeson wanted him to stay on as a tactical adviser for Fighter Control, another six months attached at Wing here at Chengtu. It was a compliment framed as a requirement, the sort of offer that acknowledged experience while quietly removing a man from his own unit.

Watts said no. He did not argue. He did not posture. He simply declined. If ordered, he would do it. Until then, he remained where he was. The possibility hovered anyway, unspoken but present, following him through briefings, across the line, and back into the quiet moments when decisions felt heavier than flight.

Colonel Cellini took the 33rd FG on June 11 and set about a practical miracle. Nearly eighty replacement pilots were extracted from Karachi, men who had been waiting in limbo while aircraft and paperwork lagged behind them. Cellini did not necissarily re-shape the 33rd, he made sure it survived. There would be a fighter group to inherit, and a future that did not depend on the endurance of a handful of veterans.

The Matterhorn Rising

Then, the scale changed.

From India's dust came an aircraft unlike anything China had seen. The B-29 Superfortress dwarfed the bombers that had preceded it. Men who had worked B-24s stopped short the first time they stood beside one. The Liberator had been a workhorse, wide-winged and familiar. The B-29 was something else entirely, taller, broader, heavier in every dimension, its engines mounted like industrial machinery rather than airplane parts. It did not look like a bomber meant to operate from improvised strips at the edge of the world.

Everything about it carried excess. More fuel. More bombs. More altitude. More demand. Where earlier bombers could be coaxed into the air with improvisation and nerve, the B-29 required mathematics, logistics, and faith in systems never tested this far from factories and ports. Its presence alone changed the tone of the field. Mechanics spoke differently around it. Pilots looked longer. The ground itself seemed too small.

Under the codename Operation MATTERHORN, the XX Bomber Command staged these giants forward onto the hardpan runways around Chengtu. Each landing was a negotiation with gravity. Fuel arrived in fifty-five-gallon drums flown piecemeal over the Hump. For every ton of bombs lifted toward Japan, nearly twenty tons of gasoline had already crossed mountains that swallowed airplanes whole. The scale was staggering and deliberate. This was no raid. It was a declaration.

On 15 and 16 June, the giants lifted east. Late departures. Heavy climbs. Bound for the Home Islands. Radios crackled. Men leaned close as fragments arrived. Up. Over target. Yawata.

Watts noted it without ceremony. "The B-29s started pulling out of here late yesterday afternoon. Quite a little flight."

Since the Doolittle Raid of April 1942, when B-25s struck Japan in a daring but symbolic blow, American aircraft had not reached the Home Islands. Now, for the first time since, American heavy bombers were launching from Chinese soil itself. Watts was not part of the operation. He was there nonetheless, moving through his own tasks while watching the crews prepare and the aircraft lift, close enough to feel the weight of it. Fighters and bombers shared the same sky, one defending a fragile perimeter, the other reaching halfway across the world.

At dawn, returning B-29s sat sweating oil into Chinese mud. Chennault called it waste. Chiang Kai-shek called it resolve. Washington called it history.

MATTERHORN revealed two truths at once. China could strike Japan. And doing so would bleed her thinner than anyone had admitted.

Only then did the long eastward drift resolve. From North Africa to Italy to India to China, the arc closed. While Europe swallowed its gamble and the Pacific felt bomb-bay doors opening over Japan, the men around Chengtu waited. Their front was quiet. Their war was stitched to the others by radio static, delayed headlines, and aircraft climbing slowly into the east.[7]

Notes

1. Charles F. Romanus and Riley Sunderland, Stilwell's Mission to China, U.S. Army in World War II: China–Burma–India Theater (Washington, D.C.: GPO, 1953).
2. Blanchard K. Watts, Handwritten Diary and AAF Form 5 Flight Logs (May–June 1944), author's collection.
3. Wesley Frank Craven and James Lea Cate, The Army Air Forces in World War II, Vol. V: The Pacific: Matterhorn to Nagasaki (Chicago: University of Chicago Press, 1953).
4. Barbara W. Tuchman, Stilwell and the American Experience in China (New York: Macmillan, 1970).
5. Samuel Eliot Morison, History of United States Naval Operations in World War II, Vol. XI (Boston: Little, Brown,

1957).

6. Stephen E. Ambrose, D-Day, June 6, 1944 (New York: Simon & Schuster, 1994).

7. Romanus and Sunderland; Craven and Cate, for MATTER-HORN operations and strategic context.

Chapter 22

The Stage and the Smoke

Restlessness spread through the 33rd Fighter Group like a fever without motion, an itch under the skin. On a foray into Chengtu proper On June 29, Watts ran into old friends from the 59th Squadron. They were staging a blowout at Fungwhanshan with wine and song, bad and good alike, the cramped joy of young men stealing hours from war. For a moment he let himself be pulled into their orbit. After weeks of mud and attrition, it felt exotic to laugh without needing a reason.

Before the party deepened into frenzy, he excused himself. "CO can't wake up drunk," he muttered, half to the courtyard, half to himself. He heading back to Pungchachung beneath paper lanterns swinging in damp night air, driving back toward a post where responsibility waited.[2]

Tragedy at the Runway

The morning of 30 June brought no mercy. Heat pressed down on the field at Pungchachung. Lieutenant Adams rolled first, his P-47 throwing up a wall of ochre dust that swallowed him whole. Lieutenant Porter followed, trusting a strip he believed was clear. The collision was decided the instant physics took over.

Watts was fifty yards away in operations. He was moving before the sound finished arriving. Adams's cockpit was crushed, flames climbing the armor. Porter's fighter flipped onto its back, the cockpit glow turning from orange to white. Crews surged forward with a weapons carrier, hooked a winch to torn metal, and hauled with a fury that ignored heat and exploding ammunition. Adams came free lifeless. At Porter's wreck men screamed that they could hear him and tried to lift the burning aircraft with bare hands. The fire answered. Both pilots were lost.

The field stank of scorched rubber and paint long after the smoke died. Watts wrote only, "Saw them burn." That night, the heat returned every time he closed his eyes.[2]

Three days later, the moment returned in writing.

On 31 July, Colonel Cellini issued a formal citation to Watts and the men who had run toward the burning aircraft instead of away from it. The language was spare, commending conspicuous courage and presence of mind under conditions that offered no margin for survival. It recorded the facts without ornament: engines aflame, ammunition cooking off, men exposed while attempting rescue.

Watts made no comment. He placed the citation into his service file and closed the cover. In China, discipline holding under fire counted as enough.

The Stage and the Smoke

China did not need the Japanese to claim its toll. On 2 July, a P-47 buckled on landing after weeks of rain softened the strip. The nose gear snapped, metal driving into mud with a sound that carried across the field. The next day another aircraft burned midstrip, men beating at the flames with extinguishers, their silhouettes cut in orange.

July's record opened and closed the same way. More losses to ground and weather than to enemy steel.[1] [2]

At one end of Kunming's runway, banners unfurled and cameras flashed. Vice President Wallace clasped Chiang's hand, Madame Chiang translated his words into song-like cadences, and photographers immortalized unity. Silk scarves fluttered; speeches leapt into cables and editorials bound for Washington.

At the other end of the strip, a gutted Warhawk still smoked where crews had dragged it into the muddy grass. Men stepped around scorched rubber and extinguished puddles. The contrast was grotesque but undeniable: statesmen posed for history while mechanics scrubbed black grime from their hands.

The Vice President's World Mission, 1944

Dispatched by Roosevelt, VP Wallace embarked on an unprecedented eighty-day diplomatic tour to reassure allies, measure adversaries, and sustain America's global image. [4]

Itinerary:

- June 1944: Alaska & Siberia, inspecting Lend-Lease ALSIB ferry routes and Soviet cooperation.
- July 1944: China, visiting Kunming and Chungking; meetings with Chiang Kai-shek and Madame Chiang, promises of allied solidarity.
- Late July–August 1944: Middle East and Mediterranean (Cairo, Jerusalem, Libya); onward to Africa and South America before returning stateside.

Reception: It was mixed at home: praised for idealism, mocked as naïve. In China, his image splashed across newspapers as visual proof the U.S. had not abandoned them. For Chennault's men, the disconnect between rhetoric and reality remained unbridgeable.

Legacy: For historians, the mission's value was symbolic more than strategic. It revealed starkly how Washington defined China as a secondary front even as it staged optics for American newspapers. [5]

Washington and Chungking spoke endurance in oratory. Kunming's endurance was an engine coaxed off wet clay, a P-47 clawing free of monsoon ground, and a squadron sent up again into weather that ate machines.

A War Slipping Away

As Wallace departed, the façade cracked in earnest. Operation Ichi-Go surged across central China, eleven Japanese divisions smashing rail lines, overrunning provinces, and scattering Chiang's armies faster than anyone in Kunming could stabilize them. Airfields evacuated ahead of advancing columns. Fuel dumps were abandoned. Promises delivered in silk banners dissolved into geography.

In Washington, the Vice President's mission would later be judged symbolic rather than strategic. On the flight line, no verdict was needed. Men watched maps redraw themselves in real time and understood what symbolism cost when measured against railheads and runways.[5]

China was not being held. It was being endured.

A Brief Command

On 18 July, the order arrived without preamble. Watts was appointed Deputy Commander of the 33rd Fighter Group, effective immediately.[1] Colonel Christman received his orders for home that same day. Major Shelton's followed close behind.

For a narrow window, authority settled where it had not been expected. The former Joker line pilot now carried responsibility across three squadrons scattered through Yunnan, hundreds of men, dozens of aircraft, and a front that was slipping east and south at once. Christman was still present in name, but the weight of execution rested with Watts.

It should have felt like arrival. His diary shows something else. His recommendation to return stateside had gone in before the others and now was left holding continuity while old friends packed. The responsibility was real, but, the timing was hollow, he'd been in active combat for the better part of 20 months with 118 sorties, he naturally wanted to go home.

The moment passed quickly. The weight did not.

The circle closed itself. One of the original Jokers became, briefly, the man holding the line.

A Polite Refusal

A mere 18 hours after Major Shelton and Colonel Christman received their orders for home, the phone rang in the operations shack. Wing was on the line. Watts's own orders had finally come through. He was going home.

He read the message with hunger and with a strange hollowness. The 180-day "loan" to China was closing out. The war would continue without him.

The mechanics of their departure fell to Watts. Christman and Shelton were already packing. BK was still half-packed when word came down that General Gilkeson had space for all three on his aircraft to Kunming. If they could get there, they could walk into Twelfth Air Force headquarters, collect written orders, and begin the long road west.

In the noise and draft of the cabin, Gilkeson leaned over and asked Watts directly what he planned to do when he returned to the States. Watts knew exactly what the general was probing. Gilkeson wanted him up at Third Air Force.

"I'd be a fool not to at least be agreeable," Watts wrote later in his diary.

He declined, as politely as a junior officer could to a senior general. Gilkeson nodded once, satisfied, and said he would help with a recommendation. "A proud moment for me," Watts noted afterward.

Get Thee to the Bank

At headquarters, collecting the orders was almost anticlimactic. They walked into the 14th AF HQ in Kunming and had the paperwork in hand within five minutes. The difficult part came next.

Watts and Shelton peeled off and headed for the finance office. Servicemen departing China were allowed to exchange up to forty American dollars into Chinese National currency, then use that currency to purchase a bank draft drawn on a New York bank, up to a ceiling of two hundred dollars.

The sequence mattered.

First, they headed over to the finance office. Watts laid forty American dollars on the counter and walked out with roughly eight thousand Chinese National dollars at the current rate. The stack of paper looked more like play money than wages, but it was the key.

Next, into Kunming proper. Two miles on foot, then a rickshaw for the last mile, clocks ticking and Christman back at headquarters unaware of where his two junior officers had disappeared to.

Somewhere in town was a Chinese bank that would issue a draft on a New York bank for the full two hundred dollars allowed by regulation. They found it. The clerk labored over a typewriter and a rubber stamp, methodical and unreadable. The result was a clean piece of paper in English and numbers, stripped of mud, weather, and improvisation. If it reached New York, it would clear for two hundred solid American dollars.

They still had to get back to the flightline.

Watts and Shelton tucked the drafts away and hurried the three miles back toward headquarters, watching the angle of the sun and the sweep of clocks in every office they passed. Christman was waiting, orders in hand, ready to move. He had no idea that in the last hours his subordinates had multiplied their pay fivefold.

They slipped back into the departure flow just in time. A quiet piece of financial engineering at the end of a long loan to China.

One last enemy waited beyond the cockpit.

At Khartoum, fever took him and malaria struck hard. Nights broke into chills. Days soaked through with sweat. "Just my luck," he wrote, "twenty-one months in combat and on the trip home was in the hospital for the first time."[2]

Familiar Faces and Empty Days

Discharged from a cot on 11 August 1944, Watts stumbled across another improbable reunion: J.P. Long, a B-25 man from the Mediterranean, appearing out of Africa's desert transit like a ghost of earlier campaigns.[3]

From Khartoum to Accra, Gold Coast, days trudged in silence. His diary collapsed to sterile notes: hours logged, nothing said. "This is my seventh day at Accra and each one getting longer," he wrote, marking endurance in boredom rather than fire. Getting his final hop home was taking longer than he could imagine. [4]

Miami: The Circle Closed

The C-46 finally settled onto the runway at Miami Army Air Field without ceremony[1]. Palm trees leaned into humid air. Cadets drilled under barked commands. Coca-Cola vendors crossed the tarmac as if the world had never narrowed to flak bursts and frost-bitten wings.

BK stepped onto American soil thinner than when he had left, fever still lurking in his bones, but upright. His flight log entries for the seven hops that brought him home were stripped to essentials: aircraft type, hours, date. No commentary.[1]

His war diary ended on 18 August 1944. Not with reflection, not with relief. It simply stopped. The ledger closed the way it had always recorded things: when the job was finished, there was nothing more to say.

Behind that quiet ending stretched TORCH, El Guettar, Pantelleria, Sicily, Salerno, Anzio, the Yellow River, and Chengtu. Three continents. Two fronts. A continuous line of responsibility carried forward without pause for the past 21 Months.

For BK, the war did not end in a dogfight or a speech. It ended in transit, in heat and waiting, in a body finally forced to stop moving. He had done what was asked of him, stayed where he was placed, and brought his men as far as orders could carry them.

Ahead lay Patterson, Muroc, and test programs where peace proved as unforgiving as combat. Pilots joked that peacetime killed almost as efficiently as war. Watts would learn that lesson next.

But that would be another chapter.

Notes

1. Blanchard K. Watts, *AAF Form 5 Flight Logs* (July–August 1944; incl. 2–3 July accidents, 18 July transfer), author's collection.
2. Blanchard K. Watts, *Handwritten War Diary* (January-August 1944; incl. 29 June party, 30 June tragedy, July fever notes), author's collection.
3. Charles F. Romanus & Riley Sunderland, *Stilwell's Mission to China*, U.S. Army in World War II: China–Burma–India Theater (Washington, D.C.: GPO, 1953).
4. Henry A. Wallace, *Mission to Asia: Diary, 1944* (Washington, D.C.: GPO; Library of Congress Manuscript Division).
5. Barbara W. Tuchman, *Stilwell and the American Experi-*

ence in China, 1911–45 (New York: Macmillan, 1970).

PART V
TEST

The warfighting odyssey was over, but peace offered no gentler sky. For Major Blanchard King Watts, the end of combat did not mean the end of risk. The calendar would turn from 1944 to 1945. The P-40 and P-47 gave way to Bobcats, Mitchells, and the first P-80s. The cockpit, however, still carried the same truth. Every flight could be final.

Warfare had demanded endurance. Test flying demanded trust. Trust in machines not yet proven. Trust in engineering drawings with the ink barely dry. At Patterson Field and later at Muroc Army Air Field, Watts entered a new crucible. This one was not marked by flak or tracer fire but by needles edging into red, by wings shuddering at the boundary between theory and failure, by cockpits where the pilot served as both operator and experiment.

Combat had taught him how to survive. Flight test would ask a harder question: Whether survival had only been preparation.

The Cold War was already gathering shape, not yet named but clearly felt. Its front line would not be beaches or valleys but altitude, speed, and metallurgy. Men who had already fought one war in the air were now asked to be the first to fly into the next one

Chapter 23

Change is Afoot

Miami was only a doorway back into the United States, not a station for BK. Within weeks of setting foot on American soil, Major Blanchard King Watts was moving again, this time north to Pennsylvania. Olmsted Field sat beside the slow slate water of the Susquehanna River, its rows of hangars hunched against a winter sky. Freight cars whispered along the river spur, air brakes squealing, steam lifting into the cold. Inside the bays, the light was hard and white, the kind that made aluminum look like bone.

Before leaving the coast, BK was routed through Philadelphia for high-altitude testing. His older brother, Lt. Cmdr. Daniel Watts, ran a Navy medical office with a doctor's unhurried authority. The clinic smelled of iodine and paper. Dan made an introduction in the corridor, and the room seemed to sharpen.

Helen J. Johnson, called Jeanne, looked up from a ledger.

Striking was the first word, not only for the dark hair and steady eyes, but for the focus behind them. Swedish family by way of Chicago, she carried herself like someone who had already solved the problem and was giving you time to catch up. They talked for five minutes that felt like twenty. Illinois Institute of Technology. University of Chicago. Hunter College training. Great Lakes. Now the Philadelphia Naval Hospital. She spoke clean Midwestern, each sentence fitted to purpose.

When BK mentioned oxygen limits and climb rates, she nodded and asked about the pressure schedule as if she had been in the cockpit beside him. Dan trusted her. BK noticed.

At Olmsted, glamour stayed outside the fence. The field belonged to the First Air Service Command, the depot lattice that kept front-line squadrons alive. Days ran on grease pencils, rivet guns, and inspection stamps. Engines coughed awake in frost while crews in wool coats leaned into prop wash. BK's Form 5 changed its voice. UC-78 hops for parts and people. AT-6 circuits to keep hands honest. B-25s with patched skins and revived hearts. Aircraft that lifted from Middletown might land next in England, then again over the Ardennes or the Rhineland. As time went on, some of the equipment would head for the Far East as well. Aside from the CBI, Europe still held his thoughts. His job was to remove doubt from machines other men would trust with their lives.[1] [2]

He learned to see the war differently. Calipers spoke as loudly as radios. A seam sealed tight against Atlantic salt could decide whether an aircraft climbed or became a name typed into a telegram. The depot was arithmetic you could hear and smell. Subtract fatigue. Replace what had failed. Return readiness to the line. The math decided battles he would never see.[1] [3]

Finishing the Job

By early 1945 the news boards filled faster than men could pin them. The Bulge broke. Patton drove east[5]. Eisenhower's front rolled over the Rhine. In the hangars, talk stayed quiet, pride was wrapped in work. An inspection tag initialed in Pennsylvania felt like a ghost hand pushing a bridgehead forward through sleet and mud.

Between flights, BK wrote letters. Jeanne's replies arrived on thin hospital stationery, her hand leaning forward across the page the way it did across a medication chart. She sent ward stories, dry and humane. She teased about Dan's exacting standards and asked about vibration notes with the attention of someone who thought in tolerances. She stayed with him in memory. Dark hair pinned neat. Chicago eyes. A mind that refused to idle.

Precision ruled at Olmsted. A bolt torqued wrong in Middletown could kill a crew over Germany. A test hop that missed a

168

shudder could doom a squadron. Different tools, same weight.[1] [3]

Yalta Summit

In February 1945, while BK ferried trainers and Mitchells out of Olmsted, the Allied leaders met at Yalta.[4] Roosevelt pressed for a peace bound by institutions and the promise of a United Nations. Churchill left there uneasy, already sensing a curtain falling across Europe. Stalin signed language about free elections and thought instead of armies. Germany would be divided. Berlin cut again within it. Poland pushed west. The Soviets pledged to enter the Pacific war. The architecture of peace was drafted even as its cracks showed.

Reactions split immediately. Some called it necessity. Others tasted surrender in the ink. Patton, never cautious even in private, said the line that would outlive him: "We defeated the wrong enemy."[5]

From Europe came photographs of Yalta. Roosevelt pale but insistent. Churchill with his pen and his frown. Stalin composed and unreadable. Men studied the captions and felt maps being trimmed with a blade finer than steel.[6]

From the Pacific came images of Iwo Jima. Ash-colored rock. A flag planted high. Casualty numbers that quieted the hangars. Victory in Europe felt inevitable. Victory against Japan still looked like blood measured in yards.

While statesmen argued over borders, BK's horizon opened onto something else. Patterson Field lay under the Ohio sky like a city of sheet metal and promise. Prototypes slept behind hangar doors. Engineers walked with slide rules in their pockets like sidearms. Somewhere, a thin, unfamiliar whine threaded the air. The war was ending, but another contest was already taking shape, and he was drifting toward its first cockpit.

Transfer Orders

On 28 February 1945, a clerk handed BK a set of orders that changed his horizon. Report to Patterson Field, Dayton, Ohio. He folded the paper once, felt the crease like a hinge, and looked out across the river where ice drifted in plates.

He wrote Jeanne that night.

See you in Dayton soon. Keep the numbers honest.

In the morning, he walked the line one last time, watched a Mitchell lift into gray light, and understood that his war was about to move from repair to invention.

Notes

1. George W. Cullum, History of the Army Air Forces Service Command (Washington, D.C.: USAF Historical Division, 1945).
2. Blanchard King Watts, AAF Form 5 Flight Logs (Aug. 1944–Feb. 1945), author's collection.
3. William Hull, The Army Air Forces Depot System in WWII (Maxwell AFB, AL: AFHRA, 1952).
4. "Mrs. Jeanne Watts Is Company's First Graduate Lady Engineer," The Spotlight (Carolina Power & Light Company employee newsletter), vol. XIV, no. 11 (May 1951), 7. Confirms Jeanne's Hunter College training; service at Great Lakes and the Philadelphia Naval Hospital; professional connection to Lt. Cmdr. Daniel Watts, USN; meeting BK; and later June 1945 Dayton marriage during his Wright Field posting.
5. George S. Patton, The Patton Papers, Vol. 2: 1940–1945, ed. Martin Blumenson (Boston: Houghton Mifflin, 1974).
6. Winston S. Churchill, "The Sinews of Peace," Westminster College, Fulton, Missouri, March 5, 1946, in The Complete Speeches of Winston S. Churchill, ed. Robert Rhodes James (New York: Bowker, 1974).

Chapter 24

Into the Test Zone

On 28 February 1945, Major Blanchard King Watts slipped a fresh set of orders into the lining of his A-2 jacket. Olmsted was finished. His destination was Patterson Field. Located outside Dayton, Ohio, it was known more formally within the Army Air Forces as Wright Field. Names shifted easily there. What did not shift was the purpose. This was where the service gathered its experience and pressed it forward, where combat record mattered less than judgment, and where no one cared where you had come from so long as you came back with the numbers right. For Watts, three campaigns older, with desert dust and Kunming clay still fresh in his memory, it felt less like a posting than an entry into a different kind of proving ground.

Wright Field was half museum, half laboratory. On one pad stood battle-worn P-40s just off depot lines; on another, crates stenciled with German eagles were pried open to reveal the shark-sleek fuselage of a Messerschmitt jet. The place smelled of oil, splintered timber, brake pads, and blueprints. Mechanics moved checklists like lifelines. Engineers stalked the hangars with slide rules in their pockets. For Watts, with the war behind him, it felt like stepping into a forge where American aviation compressed every theater he had fought in and unveiled what would come next.

A photograph from early Summer 1945 shows the cadre, some standing square to the lens, others half turned or looking away. Not all can be named, but several are marked. They were the faces of Wright's wartime transition, men standing in sunlight and shadow at the edge of the jet dawn. Also pictured in the cadre photo is P-80A-1-LO aka "44-84995", a Wright Field Flight Test Division jet used in 1945 to evaluate the feasibility of towing a fighter behind a B-29 to extend range, an interim solution for early jets with short legs before aerial refueling matured. The work emphasized a stable tow speed window, safe release procedures, and control limits to avoid towline-induced oscillations. 44-84995 later crashed near Wright Field on 23 September 1947.[1][2]

Fig 24-1 Summer 1945 Wright Field, Vandalia, Ohio. L to R: P80 #44-84995, unknown, unknown, Capt. Don Gentile, 1Lt. Robert A. "Bob" Hoover, unknown, Flight Officer Ornerlein, Capt. Don Trowbridge, Unknown, Maj. John T. Godfrey, 1LT Joseph Grabowski. Annotations by BK Watts (Author's collection)

The Wright Cadre

The photograph was taken at Wright Field during the first uncertain months of the jet transition. Names were later written along its margins, some familiar, others barely recorded elsewhere. Together they form a cross section of the pilots moving through Wright at the moment when combat aviation gave way to experimentation.

Captain Don S. Gentile, identified on the Wright Field order by service number O-885109, was the decorated World War II ace Dominic "Don" Gentile. In this photograph he appears not as a celebrity, but as a working officer among peers, one of the combat-experienced pilots assigned to Wright Field during the transition from wartime operations to jet testing.

Bob Hoover is listed by name as well. His wartime escape from Europe and later career would make him legendary, but at Wright he was one of several pilots flying captured German aircraft and early American jets, including the P-80. Even then, those around him recognized the combination of discipline and instinct that later defined his reputation as a pilot's pilot.

Ornerlein's name survives almost entirely through this image and Watts's own notes. He flew alongside the better known figures, strapping into the same unfamiliar aircraft, including German types brought back for evaluation and the first operational jets. His work was hazardous and largely uncelebrated, representative of the many pilots whose contribution never reached formal histories.

Don Trowbridge appears in the same frame, another of the professionals who carried the burden of proving systems rather than chasing distinction. His flights generated data and confidence, the quiet foundation on which others built.

The photograph also records Major John T. Godfrey, recently repatriated from combat with the 336th Fighter Squadron. His presence reflects the brief overlap when returning aces passed through Wright alongside officers shaping the postwar air force and he had flown in Europe with ace Don Gentile during the air war over Germany.

Joseph Grabowski, listed as a lieutenant, represents the younger edge of the cadre. He was learning jets as they were being written into logbooks, absorbing risk without fanfare, a junior member of a fraternity still defining itself.

Seen together, the photograph does not document a formal

unit or program. It captures a moment. Combat veterans, test pilots, and newcomers shared the same ramps and hangars, bound not by fame but by proximity to machines that no longer tolerated improvisation.

Watts stood among them, not yet singled out, part of a working cohort at the threshold of a new kind of flight. In the photograph, alongside those named and annotated by BK, stood others he did not identify. It is tempting to cast them as test pilot heroes, and in most cases that instinct is probably correct. This was a moment when the men in that photo helped decide, for themselves and for the United States, what the Air Force would soon become.BK spoke little about this group in later years. What remains are documents, flight logs, and a handful of photographs. The unnamed pilots in this image stay unidentified, but together they represent an Air Force learning to walk in the jet age while still unbuckling the last war.

Reasons

The jet age, however, did not wait politely for the air force to catch up. Even before Watts's orders sent him west to Muroc, the P-80 program had already drawn blood. The second production YP-80A, tail number 44-83024, had been stripped of its guns and converted to a photo-reconnaissance ship, re-designated XF-14. On 6 December 1944, during a test flight near Muroc, the camera-nose P-80 collided in midair with a Lockheed-owned B-25J. Both aircraft came apart in the same clear desert sky. Pilot Perry B Claypool, flying the P80 and everyone aboard the bomber, piloted by Henry M. Phillips, died.[8]

In Watts's surviving photo files, a small series of black-and-white prints almost certainly captures that airplane and pilot alive before their fate caught up. Right and left oblique passes, a full starboard view from a chase plane, and a near-overhead plan-form, all over the familiar hard light of the Muroc range. The fronts of the photos carry captions with file numbers in the same tight range and style as other Air Force photographs from his Muroc and Wright days. The other photos, sharing markings and showing the same airframe from different angles, are treated as additional views from the same or closely related flights. What was routine documentation in 1944 reads now like an unintended storyboard of a jet that would soon vanish into an accident report.

Fig 24-2 Undated photo of YP80 Tail number 483024. Reverse of photo has a serial number "100771", consistent with the numbering style of other photos in the Author's collection

What Was Really Happening

At Wright Field especially, flight test meant more than top end speed runs and record attempts. It was the disciplined business of proving the entire system, strategy meshing with tactics, and tactics meshing with machines. The first big postwar jet problem was range: the P-80 could fight, but it could not loiter or roam without burning its radius on the way to the fight. Towing a jet behind a Superfortress sounded inelegant, but it was a practical bridge to the future, a way to move combat power without spending it en route. The tasking was as unglamorous as it was essential: find the tow speed band that stayed stable in real air; define a release that did not threaten the engine; write the turn limits that kept the towline from walking toward the tail; and codify the "knock it off" calls. These were the abort calls that ended a run the moment oscillations went from lazy to dangerous.[1][2][5]

This was what Flight Test did day after day. A test card might read like a grocery list, airspeed steps in five knot increments, cowl and canopy settings, bank angles capped at fifteen degrees, yet the results wrote doctrine. A confirmed tow speed window became a training standard; a workable release procedure became a checklist item; a rudder trim bias under tow saved pilots from chasing oscillations when the workload spiked. Even the maintenance notes mattered. Tow hardware stressed attach points and wiring runs; inspections and reinforcement schedules followed. The work looked mundane from the ramp. It was not. It was the team engineering of risk out of the system. [3][4][5]

For the men in the cockpit, that quiet engineering lived alongside something more primitive. Pilots felt the line begin to hum, the nose start to hunt, and some animal part of them wanted to fix it with muscle. The test pilot's job was to feel that urge, note it, then stay inside the card. Resist the impulse to improvise until the "knock it off" call, then fly the abort as cleanly as the planned run.

If the numbers were modest, the stakes were not. Proving that a jet could be delivered under tow to a release point with cooling margins intact and control authority in hand meant commanders could plan beyond the map edge before tankers and forward basing caught up with strategy. When a trial exposed a limit in turbulence thresholds, or in bank angles that stacked drag faster than rudder, those red lines saved crews who would never know their lives had already been protected by someone else's boring test flight.

The culture at Wright prized this kind of clarity. A successful sortie was one that ended with fewer unknowns than it began, data that closed a loop between the airplane as it was advertised and the airplane as it was flown. 44-84995 stands in that lineage. Its role in the B-29 tow feasibility work is a reminder that progress is often incremental, procedural, and uncinematic, built on flights whose drama lives in the margins of a test card. The result was an Air Force that could think in systems, strategy, tactics, and equipment aligned, not just in speed and altitude.

That systems thinking permeated everything on the line. Tugs moved airframes not because a pilot could not taxi, but because nose struts had life limits and brake temperatures wrote their own failure stories. Tow bars, wing walkers, chocks placed and pulled on cadence, each action had an accident report behind it and a checklist ahead of it. A tow at ten knots on dry pavement taught something about shimmy dampers; a tow in a quartering tailwind taught something about gust loads and fingertip aileron inputs even when engines were silent. Maintenance chiefs could quote the paragraph where a tow fitting had been re-stressed; pilots could tell you the lateral stick displacement that kept a jet's nose from hunting behind a tug. Nothing was wasted. All of it was data inside the paradigm of test flight.[3][4][5]

Briefings reflected the same economy. Objectives were simple and numbered. Hazards were named. The abort criteria were memorized. The test card was king. You went to the airplane understanding what one knot too fast could do to a towline oscillation or how one degree too much bank could load the line beyond safety margins. You came back and wrote down exactly what the air had told you. Someone else's life would depend on your decimals. This was what it was all about.

Transition

BK's AAF Form 5 reflected the change immediately.[3,4] The old combat diary of sorties and dogfights gave way to a working ledger:

- UC-78 Bobcats, shuttling personnel and parts over the quilt of Midwestern farmland.
- AT-6 Texans, avionics and proficiency hops, the same trainer that had taught his first tricks now serving as his metronome.
- B-25 Mitchells, once Tokyo raiders, now test mules burdened with clipboards and engineers strapped beside him in the jump seat.

Each line in the log captured not drama but diligence, a new kind of warfare measured not in enemy losses but in tolerance stacks and vibration notes.

Dawn of the Jet Age

By April 1945, with a quick stint at Muroc, Watts' AAF Form 5 no longer wavered. Day after day, line after line, it read the same: P-80 Shooting Star.[3][4] In California, Watts left testing the piston age behind. Rogers Dry Lake spread beneath him, a sun bleached plain hammered flat for millennia, perfect for America's jet baptism. For two months he flew the Shooting Star almost exclusively, strapping into a future that only weeks before had been the stuff of rumor.

The sensations were uncanny: no prop wash, no piston tremor. Only a turbine's thin whine swelling into a searing hiss, acceleration that built like a held breath, thrust as smooth as it was relentless. Watts, combat hardened by nearly two years of combat, became something else here: a jet pilot at the dawn of the age. The desert no longer tested survival by flak but by physics, and survival was no less tenuous.

Fig 24-3 USAF photo, April 1945. XP-80s on North Base flightline, Muroc AAF (later Edwards AFB). Courtesy Ken G. Oldfield. Library of Congress, HAER CAL,15-BORON.V,2--6. Public Domain.

Then, almost abruptly, orders pulled him east again. His log swung back to Wright Field chores: AT-11 Kansans for navigation trials, B-17s cut for radar noses, C-47s hauling freight and men between depots.[3] [5] Jets whispered at the edge of memory, but the Midwest skies filled with the sounds of round engines, props, and the weight of America's arsenal being inventoried one final time.

Victory and an Unfinished Future

In April 1945, the radios crackled with news that stunned the country. Franklin Delano Roosevelt was dead. A fourth-term president carried away during a Georgia spring, leaving the nation mid stride. At Wright Field, newspapers lay open in hangar corners, black borders framing the loss. For pilots raised on Roosevelt's voice since cadet days, it felt as if the keel had been knocked from the ship.

A month later, on 8 May, the guns fell silent in Europe. V.E. Day brought church bells and brief celebration. Watts thought of George S. Patton, the armored storm who had once stood before them in North Africa, driving men forward with the force of artillery. Somewhere in occupied Bavaria, Patton's columns were still moving when the ceasefire froze them in place. Even at Wright, the memory of his general's voice lingered, sharp as a cadence call.[7]

Then, in August, came Hiroshima and Nagasaki. The war ended not with a final advance, but with a rupture so sudden it unsettled even its victors. Peace arrived whole, unfamiliar, and heavy with consequence.

Dayton did not erupt the way front-line fields had. There was no battered flight line to gather on, no squadron bar to crowd into. Victory looked like engineers sliding papers across drafting tables and pilots flying Bobcats and Mitchells on depot checks while parades passed beyond the gate. The Pacific still burned, and euphoria kept its distance. [5] [6] [7]

For the test pilots at Wright and Muroc, the lesson was brutal: the age of conventional war was over. The next conflict would be fought with weapons and machines unimagined just three years before.[8]

Watts felt it with every memory of turbines singing over Rogers Dry Lake. He had fought in three theaters, saluted Patton, mourned Roosevelt, and now watched the world split open under atomic light.

The war was finished.

The desert was just beginning.

Notes

1. Joe Baugher, "USAF Serial Number Search Results: 44-84992 to 44-85336," accessed November 7, 2025, http://www.joebaugher.com/usaf_serials/1944_4.html.
2. "Forgotten Jets – Lockheed P-80A 44-84995," accessed November 7, 2025, http://www.forgottenjets.warbirdinformationexchange.org/.
3. Blanchard King Watts, AAF Form 5 Flight Logs, January to September 1945, author's collection.
4. William Hull, The Army Air Forces Depot and Test System in WWII (Maxwell Air Force Base, AL: AFHRA Historical Studies, 1952).
5. Doris Kearns Goodwin, No Ordinary Time: Franklin and Eleanor Roosevelt: The Home Front in World War II (New York: Simon and Schuster, 1994), 612–618.
6. Carlo D'Este, Patton: A Genius for War (New York: HarperCollins, 1995), 788–795.
7. Richard Rhodes, The Making of the Atomic Bomb (New York: Simon and Schuster, 1986), 710–735.
8. "Lockheed YP-80A Shooting Star," Joe Baugher, accessed November 7, 2025, https://www.joebaugher.com/usaf_fighters/p80_3.html. The serial summary records 44-83024, originally the second YP-80A, as converted to photo-reconnaissance configuration and redesignated XF-14, and notes that it was "destroyed in mid-air collision with B-25J 44-29120 near Muroc AB Dec 6, 1944. All crew on both planes killed."

CHAPTER 25

THE DESERT CRUCIBLE

On 21 September 1945, Headquarters at Wright Field issued Letter Orders No. A9-21-27 for an "accelerated service test on five P-80 airplanes."[1] The paper was plain. The implication was not. These were not exploratory flights or publicity hops. They were instructions to prove, quickly and without loss, whether America's first operational jet could be trusted with pilots' lives.

The orders came from the top of the newly reorganized Air Technical Service Command, signed by its commander, Major General Hugh J. Knerr. Knerr had been placed at Wright to consolidate the Army Air Forces' technical future, to take the lessons of war and harden them into machines that could survive peace. His signature made the priority unmistakable. The P-80 program had moved beyond experiment. It was under accelerated service test.[1]

The list of pilots named on the order was short and deliberate. Eight officers. Five jets. The names appeared in rank order, with Major Blanchard K. Watts listed first. At Wright and Muroc, that mattered. Rank carried responsibility. Experience carried weight. Reputation, alone, did not.

By Summer of 1945, the jet age had erased old divisions. It no longer mattered whether a man had come up through North Africa, the Eighth Air Force, or the training command. It mattered whether he could read a test card, hold discipline under uncertainty, and bring an unproven aircraft home intact. At Wright Field and on the dry lakes of the Mojave, survival was no longer about enemy fire. It was about judgment.

The desert would decide the rest.

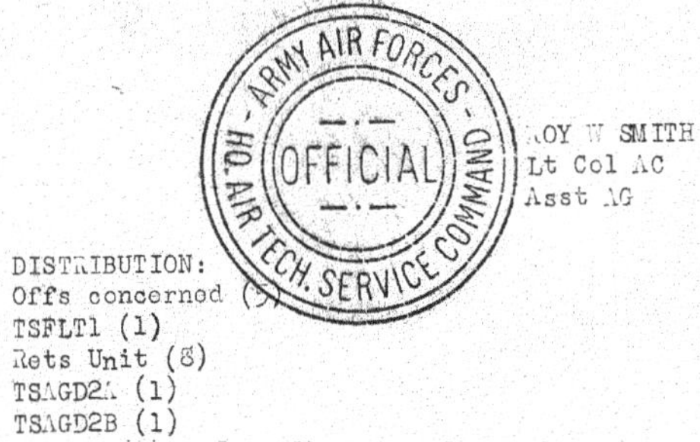

Fig 25-1 Letter Orders directing the Cadre of 8 Pilots to Muroc 21 SEP 45 to test 5 P80's (Author's Collection)

Thacker and the Five Jets

The concern behind Knerr's order was immediate and lethal. By Fall 1945, the Lockheed P-80 Shooting Star had earned a reputation as dangerous to its own pilots. Accidents came faster than explanations. Lockheed's chief test pilot, Milo Burcham, had been killed in October 1944. America's leading ace, Major Richard Bong, died in another P-80 crash on 6 August 1945[5]. The jet promised the future, but it was bleeding the present.

At Wright Field, Knerr turned to Colonel Robert E. Thacker.

Thacker was not chosen for flair. He was chosen because he finished things. In later oral history, he recalled being called in and handed the problem without ornament. Go to Lockheed. Collect five P-80s. Take them to the dry lakes at Muroc. Fly them hard. Fly them correctly. And do not lose another man.

"We went up on the dry lakes… called Muroc Army Air Base," Thacker remembered.[1] He took Captain Charles E. "Chuck" Yeager and two other pilots to Burbank, accepted five Shooting Stars from Lockheed, and ferried them east into the Mojave. Those same aircraft now appeared, stripped of story and urgency, as "five P-80 airplanes" on Knerr's 21 September orders.

The paperwork called it an accelerated service test. Thacker put it more plainly.

Five jets.

Five hundred hours.

Zero losses.

The names on the 21 September order read like a roll call of that judgment:

- Maj. Blanchard K. Watts: tested in three theaters of war, P-80 experience at Muroc

- Capt. Kenneth O. Chilstrom: methodical, destined to shape the Test Pilot School.

- Captain Don S. Gentile: the decorated World War II ace and one of the highest-scoring American fighter pilots of the European war.

- Capt. Martin L. Smith Jr.: understated, part of the younger cadre of professionals.

- Capt. John E. Sullivan: steady evaluator, another combat-seasoned addition to the jet cadre.

- Capt. Charles E. "Chuck" Yeager: Mustang ace, future supersonic pioneer.

- Lt. Robert A. Baird: junior but sharp, absorbing jet habits as fast as they were written.
- Lt. Henry A. Johnson: another of the next generation, quiet professionalism matched to endurance.

Eight names. Five jets. A deliberate blend of wartime experience and experimental discipline. Knerr's order placed combat aces beside emerging test specialists and younger evaluation pilots, men chosen not for publicity but for the steadiness required to keep a dangerous aircraft alive long enough to understand it.

They converged on Rogers Dry Lake, turbines shrieking across the Mojave. Glen Edwards, a bomber test pilot, noted in his diary on September 26, 1945, that he found "Chilstrom, Sullivan, Watts, Baird, and a few more of the Fighter Test boys are there for some P-80 work" at Muroc, confirming the presence of this elite group.[8]

The logic behind the roster was as sharp as the desert horizon. With Bong gone, the Air Forces still had exceptional pilots to call on, and they did so deliberately. These were men chosen for judgment rather than headlines. Yeager brought instincts honed in Mustang combat that would soon carry him beyond conventional limits. Others arrived with quieter reputations, proven by consistency and survival. Watts's Mediterranean record alone, with more than 110 combat sorties and three confirmed victories, marked him as thoroughly tested. He was not selected for notoriety. He was there because he finished missions, stayed inside discipline, and came back with answers.

Around them, Knerr and Thacker built out a cadre of methodical professionals and hard-nerved younger pilots. They were not chosen at random. This was the breed Tom Wolfe[4] would later say had "the right stuff": men who believed they could fix and fly anything. The personality type that needed to understand machines from their very foundation. Those who could absorb orders, take ownership of the plane, stay inside the test card, and still accept that each sortie might be their last. BK had already been at Muroc in late April and May of that year, logging twenty hours in the P-80 when most officers had never touched a turbine.[2] Thacker knew who could fly jets, who had fought and survived, and who could be trusted not to improvise themselves into the ground.

Three days after Knerr signed A9-21-27, BK's AAF Form 5 recorded a 9 hour and 45 minute flight in an XC-97 on 24 September 1945, an unusually long leg that fits the movement of the test detachment west from Wright toward California.[2]

From there, Thacker's recollection picks up the lived detail. At Muroc he stood the pilots down in a briefing and laid out the rules: stay within a dry-lake recovery box at all times. Keep enough altitude to glide dead-stick to one of the dozens of alkali pans below if the turbine flamed out. No excursions toward civil airways, and above all, no accidents. When Yeager wandered all the way up toward San Francisco the next day into controlled airspace, Thacker grounded him for two days.[1] Then, the cadre went back to work. Five P-80s, 500 hours, and not a single hull lost. That was the standard that Gen. Knerr and Col. Thacker intended to meet.

The Fraternity

Muroc in 1945 was dust, wind, and improvisation. Rogers Dry Lake stretched for thirty square miles of flat alkali, often described as "nature's perfect runway."[3] Mess halls fried beef, corrugated hangars baked under midday sun, and blowing grit worked its way into goggles and engines alike. BK had been there earlier that year, but this time, under Knerr's orders and Thacker's discipline, it felt different. The desert was no longer just a proving ground. It was the laboratory where the P-80 would either earn its place or be written off as too dangerous to trust.

By night, the fraternity gathered at Pancho Barnes's Happy Bottom Riding Club, later immortalized in The Right Stuff.[4] Bravado and whiskey mixed with dust, poker chips slapped against scarred tables, and Pancho herself laughed loudest of all, calling the jet pilots "my boys." Yeager leaned back, drawl easy. Chilstrom listened, notebook in his breast pocket. Watts, just as cocky with his North Carolina drawl, kept the liquor flowing steadily, his humor dry, and his presence sure. They were the mix Knerr and Thacker had engineered. Headline aces, quiet technicians, and combat-tempered hands like BK who knew how thin luck could run.

Beneath the talk, there was a quieter common thread. Most of them were, at heart, engineers who happened to be flyers who argued over data as easily as they traded jabs at the bar. Test cards, fuel flows, temperature limits: these were as natural a language as boasts about G-loads. BK also kept a small notebook in his breast pocket, the habit he would carry for the rest of his life. Numbers, sketches, stray ideas went into it, even shopping lists. In that way, he was typical of the fraternity. Their swagger rested on something harder than nerve. They were doers who measured, adjusted, and tried again until the machine and the man both worked. In every era, it is that kind of personality that furthers progress.

Fig 25-2 Capt John Sullivan in his P-80 Cockpit, Muroc Field September 1945. Annotations by BK Watts (Author's Collection)

Muroc from the Air and Ground

From 10,000 feet, Rogers Dry Lake shimmered like a polished mirror, its pale surface reflecting sky in faint silver. Makeshift "runways" fanned in every direction. From that altitude it seemed limitless, framed only by the distant Sierra Nevada mountains. Watts, peering from a P-80 canopy, thought it resembled a canvas offered by the earth itself for aviators to paint upon.

From the ground, the majesty shrank to raw discomfort. Sand scoured skin, winds snapped tent flaps, heat by day scorched, nights froze. Trucks sank on soft margins of the lakebed. Men spat grit, joking that "this dust forms half our rations." Engines wheezed under alkali.

Yet here, beneath hardship, they sketched the future. Two years later, Yeager would break the sound barrier above this lake. But in 1945, it was Watts and this cadre who laid the foundation, their five jets carving careful trails into the void, tethered to an ancient lake doubling as laboratory and safety net.

Fig 25-3 Unknown Pilot inverted over the Grand Canyon October 1945 Annotations by BK Watts (Author's Collection)

Bravado and Shadows

Service test at Muroc was no parade. Stopwatch climbs, controlled dives into compressibility, endless reliability checks. Watts logged the work in taut lines of neat script, hours seldom stretching past sixty. But between those lines germinated legends: Yeager adjusted instinct honed on Mustangs into jets. Chilstrom built the trade into a profession, exacting and methodical.

Watts endured the work, cocky and skilled enough to be in their company, less celebrated but combat-tempered, essential as the stars around the headline names. Now and then his camera came out, trying to catch what could not be staged. One surviving frame, grainy and off in the distance, shows a jet inverted over a gouged horizon he labeled simply "Grand Canyon 45," a moment of unscripted flight that hints at how thin the line between work and dare could feel.

From dust and hardship, the desert bred the mystique of test pilots, a fraternity destined to stand beside astronauts as America's knights of the twentieth century.

The orders had been carbon sheets. The reality was turbines screaming across an alkali lake, young pilots and veterans strapping into unproven airframes with test cards half-guessed and safety nets barely there.

Fig 25-4 Air Materiel Command Test Pilot Certificate
Official recognition of Blanchard King Watts as a test pilot in the Operations Sub-Division, FLight Test Wright Field. Signed by Col. Albert C. Boyd, Chief of Flight Test, Air Materiel Command. (Author's collection).

Choosing the Steadier Sky

By October 1945 the numbers were finally on their side. The five P-80s assigned to the accelerated service test had accumulated their hours without losing a man or a machine. Five hundred hours flown. Limits found, written, respected. The desert had not claimed them.

That mattered. Not because it ended the danger, but because it proved that discipline could bend risk into something survivable. The Shooting Star could be flown safely if the pilot stayed inside the work, trusted the card, and resisted the old combat instinct to force an answer out of the airplane. This was the lesson Thacker wanted. This was the lesson Knerr also needed carried back to Wright Field.

Men began to cycle out. Others moved on to new assignments, new programs, new risks. The fraternity loosened without ceremony, the way wartime units always did when the work was finished.

Watts logged his last Muroc sorties the same way he logged everything else. Clean lines. No flourish.[2] The desert had given him what it was meant to give. He had flown with the giants, tasted the edge of the jet age, and helped turn a dangerous promise into something commanders could plan with.

When his orders sent him back east, he did not argue. Muroc was a summit, not a residence. At Wright Field, the work continued, relentless and necessary. Depot evaluations. Systems testing. The long, patient labor of turning combat lessons into machines that would not kill their crews in peacetime.

Another pull was already stronger. The education he had left unfinished in North Carolina. The discipline to match experience with theory. Leadership that rested on more than nerve and luck.

He turned east toward Wright Field and classrooms instead of alkali and bravado. The choice was not retreat. It was selection. The desert had taught him how thin the margin was. The steadier sky was where he intended to build something that lasted.

Fraternity and Family

Even in late December 1945, months after their 60-day stint at Muroc, the brotherhood was still linked. Major Glen W. Edwards, older, steady, and flying everything handed to him, was a constant presence. Ken Chilstrom remained sharp-edged, the rules-keeper. Watts, newly married to Jeanne Johnson after a year of courtship, bridged fraternity and family. At Christmas, the three threads intertwined. Edwards and Chilstrom came east, and instead of Pancho's bar back in the desert, they dressed up for a night out in Chicago. War bonds and flight pay bought linen tablecloths, low lights, and a small nightclub table where glasses caught the glow of the bandstand.

They were not yet legends or names on buildings. They were just pilots on leave, savoring a brief taste of the good life and trying, in their own way, to build a safer, better Air Force than the one they had inherited. The future would make some of them famous. Edwards would be gone within two years, killed at Muroc in June 1947 while co-piloting the YB-49 Flying Wing.[5] The entire base would later be renamed Edwards AFB in his honor.[5] The same place where Yeager would break the sound barrier above the same desert, Chilstrom would help formalize the test-pilot trade.[9]

Photo L to R: Glen Edwards, unknown, Kenneth Chilstrom, BK Watts, Jeanne Watts. Chicago, IL Christmas 1945 Annotations by BK Watts. (Author's Collection)

The Chicago nightclub photo holds its poignancy because none of that is visible yet. In that frozen moment, they are simply part of a young fraternity...men and women at a small table, smiling into a camera, unaware how quickly history would close over their easy laughter.

Notes

1. Letter Orders No. A9-21-27, HQ Air Technical Service Command, Wright Field, 21 Sept. 1945: author's collection. The order, issued by Maj. Gen. Hugh J. Knerr, directs an "accelerated service test on five P-80 airplanes" and names Maj. Blanchard K. Watts, Capt. Charles E. Yeager, Capt. Dominic S. Gentile, Capt. Kenneth O. Chilstrom, Capt. John E. Sullivan, Capt. Martin L. Smith Jr., 1st Lt. Robert A. Baird, and 1st Lt. Henry A. Johnson. Col. Robert E. Thacker, USAF (Ret.), oral history interview, 52:32, video, posted 2 March 2018, YouTube, "Col. Robert E. Thacker Oral History," https://www.youtube.com/watch?v=YW-ow6lfYIlQ (relevant segment c. 44:30–47:30). In this segment Thacker recalls being tasked at Wright Field to take Capt. Charles E. Yeager and two other pilots to Lockheed, collect five P-80s, ferry them to "the dry lakes.... called Muroc Army Air Base," and fly them for 500 accident-free hours. On the basis of the matching number of aircraft ("five P-80s"), the 21 Sept. 1945 date of Letter Orders A9-21-27, and the 9:45 XC-97 flight logged by Maj. Blanchard K. Watts on 24 Sept. 1945 (AAF Form 5, author's collection), this book treats Thacker's "five P-80s" and the "five P-80 airplanes" in A9-21-27 as the same detachment.

2. Blanchard King Watts, AAF Form 5 Flight Logs, entries for April–May 1945 and 24 Sept. 1945, author's collection. The spring 1945 pages show Watts flying P-80 sorties from Muroc Army Air Base. The 24 Sept. 1945 entry records a 9:45 flight in an XC-97, consistent with ferry/support movement from Wright Field to California for the P-80 acceler-

ated service test detachment named in A9-21-27.

3. National Park Service, By Aviation: From Sand Dunes to Sonic Booms (Edwards AFB & Rogers Dry Lake history), retrieved 14 Sept 2025.

4. Tom Wolfe, *The Right Stuff* (New York: Farrar, Straus & Giroux, 1979).

5. National Park Service, By Aviation: From Sand Dunes to Sonic Booms (Edwards AFB & Rogers Dry Lake history), retrieved 14 Sept 2025.

6. "Milo Burcham, Lockheed Test Pilot, Killed," New York Times, Oct. 19, 1944; contemporary coverage of Maj. Richard I. Bong's death in a P-80 crash at Burbank in the Los Angeles Times and New York Times, Aug. 7, 1945. For discussion of early P-80 accident experience and safety concerns, see Richard P. Hallion, On the Frontier: Flight Research at Dryden, 1946–1981 (Washington, DC: NASA SP-4303, 1984).

7. "Maj. Glen W. Edwards Killed in YB-49 Crash," Los Angeles Times, June 6, 1947; also New York Times, June 7, 1947.

8. Glen Edwards, The Diary of a Bomber Pilot, with Daniel Ford (Washington, DC: Smithsonian Institution Press, 1998), 93.

9. For Yeager's X-1 flight at Muroc and the evolution of test-pilot practice at Wright Field and Rogers Dry Lake, see Wolfe, The Right Stuff; Hallion, On the Frontier; and National Park Service, From Sand Dunes to Sonic Booms. Chilstrom's participation in early P-80 accelerated service testing is documented in Letter Orders A9-21-27

PART VI
THE COLD WAR

The war had ended, the medals were pinned, and, for a time it seemed the cockpit could finally be set aside. Four years of peace gave BK and Jeanne Watts room to build something durable: classrooms and drafting boards, trailer walls, long Carolina roads, and the steady hum of small propellers over pine forests. It was a life assembled deliberately, one piece at a time.

History, however, rarely releases those it still has use for.

By 1951, peace was already thinning. Korea burned. The Cold War hardened from theory into posture. The United States Air Force would reach again for men who had already learned how to endure uncertainty at speed. For Watts, the summons was not simply a return to uniform, but a return to a way of being. Back into classrooms with chalk and schematics. Back onto tarmacs and flight lines where decisions still carried weight.

The intermission was over.
The next act had already begun.

CHAPTER 26
BETWEEN WARS

To you who answered the call of your country and served in its Armed Forces to bring about the total defeat of the enemy, I extend the heartfelt thanks of a grateful Nation. As one of the Nation's finest, you undertook the most severe task one can be called upon to perform. Because you demonstrated the fortitude, resourcefulness and calm judgment necessary to carry out that task, we now look to you for leadership and example in further exalting our country in peace.

Fig 26-1 Letter from President Harry S. Truman, Spring 1946
Postwar form letter to American service members calling on them to lead by
example in peace. Addressed to Blanchard King Watts as a lieutenant colonel,
dating the letter to after his promotion in February 1946. (Author's collection).

February 1946. The war wound down with parades, flags, and relief across the States, but for Lt. Colonel Blanchard King Watts the pivot from cockpit to sidewalk was quieter, almost disorienting. He stepped from the Army Air Forces after four years of constant motion across North Africa, Italy, Asia, Ohio, and California, into Winston-Salem. These were the first uncertain months of his civilian life.

He was still a combat flier at heart, yet suddenly married, homebound, and waiting for the birth of his first child. The headlines spoke of atomic secrets, Soviet advances, and an iron curtain drawing across Europe, but Watts faced a more immediate question: how to build a stable life out of peacetime's stillness.[1]

Charleston and The Citadel

By autumn 1946 their first daughter had arrived, and with her, a new resolve. Watts enrolled at The Citadel in Charleston, not to relive cadet rituals, but to rebuild his future through mathematics, physics, and military discipline applied to civilian purpose.[2]

The family settled into a twenty-two-foot aluminum trailer parked within sight of The Citadel's fortress walls. Jeanne was later quoted that it had "most of the comforts of home, if not the space."[3] Inside, Watts pored over textbooks by night; Jeanne, the former WAVES pharmacist's mate, raised their daughter on the narrow strip of linoleum that served as a floor and a playground. Outside, cadets drilled beneath the flag; inside, family and study formed a different kind of duty.

NC State: The Engineering Years

In 1948, the trailer rattled its way north to Raleigh. Both Jeanne and BK enrolled at North Carolina State College's School of Engineering. For him it was academic redemption, translating the physics of flight and the logic of maintenance into equations and blueprints. His combat logbook gave way to a ledger of problem sets and grades.

"Hard work, close budgeting of time and money, and living in a trailer!" Jeanne recalled years later with a wry smile.[4] They lived among a generation of veterans who traded flight suits for drafting tables, the air full of ambition and solder smoke. Watts brought a field mechanic's pragmatism to classroom aerodynamics. Jeanne brought something rarer still: she was one of the few women in the program, her WAVES laboratory experience now counted as scientific credit and social defiance alike.

For their daughter, the trailer park became a community nursery of shared toys and borrowed strollers. Graduation that spring meant more than a diploma. As Jeanne put it: "it meant the end of trailer living." [3]

With his degree complete, BK accepted a civilian flying job with the North Carolina State Wildlife Commission, piloting low, slow patrols over forest and river. He counted deer, tracked waterways, and traced the blue veins of the Piedmont. The cockpit that had once symbolized survival in war now framed peace and stewardship. [6]

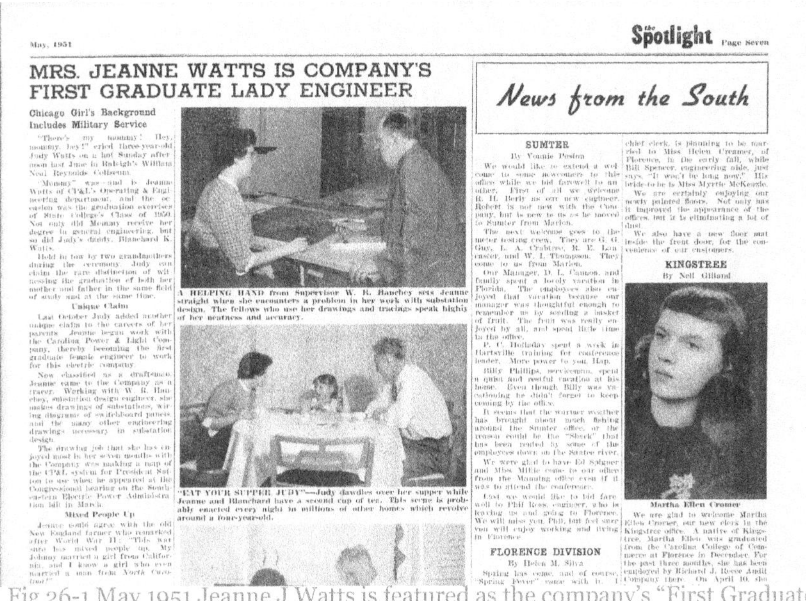

Fig 26-1 May 1951 Jeanne J Watts is featured as the company's "First Graduate Lady Engineer" in "The Spotlight", Carolina Power and Light's monthly employee newspaper.(Author's Collection)

First Lady Engineer

While BK returned to flight, Jeanne pushed forward into the professional unknown. In 1950 Carolina Power & Light Company hired her as its first woman engineer. [7]

"Mrs. Jeanne Watts Is Company's First Graduate Lady Engineer."[8] In the photo she stands at her drafting board, stylus poised, hair pinned neatly, a portrait of postwar modernity. Her job was neither ceremonial nor decorative. She calculated load potential for power lines, sketched wiring plans for rural grids, and helped push electrical service into still-dark corners of the state.

By 1951 Raleigh had become home. The Watts family left the trailer behind for a modest house near St. Mary's Street, with rooms, a yard, and Jeanne's drafting board framed against a window, a symbol of progress in both career and equality.

Jeanne Watts, A Pioneering Path

- WAVES Service (1943–1945): Pharmacist's Mate; altitude physiology studies, Philadelphia Naval Hospital.

- Engineering Education (1948–1950): Among the first women in NC State's School of Engineering.

- Carolina Power & Light (1950–1951): First female engineer employed by the company; rural electrification projects.

- Legacy: Balanced professional achievement and motherhood in an era that doubted both could coexist 4 children and 3 grandchildren later, she could have silenced any doubter.

The Call Returns

Their honeymoon did not last. On May 10, 1951, orders recalled Lieutenant Colonel Watts to active duty.[9] The Korean War demanded pilots who could bridge eras, men fluent in piston aircraft yet adaptable to jets. Watts, with combat tours, test pilot experience, and an engineering degree, fit the mold perfectly.

On April 27 Jeanne resigned her position at Carolina Power & Light.[10] She packed their household goods and their young daughter and followed her husband to Bergstrom Air Force Base, Texas. The house in Raleigh, the newsletter headline, and the rare title of "first lady engineer" folded into history like an archived clipping.

Peacetime had given them a brief intermission. The curtain was rising again.

Notes

1. "Mrs. Jeanne Watts Is Company's First Graduate Lady Engineer" The Spotlight, vol. XIV, no. 11 (Carolina Power & Light Company employee newsletter, edited by Watt Huntley), May 1951.
2. Ibid., "moved to Charleston so that Blanchard could attend The Citadel."
3. Ibid., trailer living details.
4. Ibid., "characterized by hard work, close budgeting of time and money, and living in a trailer."
5. Ibid., "Blanchard flies a patrol plane for the State Wildlife Commission."
6. Ibid., "first lady engineer ever employed by Carolina Power & Light Company."
7. Ibid., "Uncle Sam has recalled Blanchard Watts to the Army Air Corps. He is to report May 10 to Bergstrom Air Force Base."
8. Ibid., "His wife, Jeanne, our first lady engineer, left the Company April 27 to join her husband in Texas."

CHAPTER 27

COLD WAR PROFESSIONAL

In April 1951, the Air Force recalled BK to active duty at Bergstrom Air Force Base, Texas, now under Eighth Air Force with the 522nd Fighter Squadron. Newly promoted to Lieutenant Colonel, the move marked a permanent transition for him and his family, from civilian work and trailer living into the institutional rhythms and relocations of postwar military life.[1]

On the ramp at Bergstrom, the aircraft were not the glamorous jets featured on Air Force Magazine covers, but rows of North American T-6 Texans. These yellow-painted trainers, dependable and rugged, were designed for men learning the fundamentals. They represented continuity amid rapid change. Every logbook entry mattered: instrument check rides, proficiency circuits, and steady instruction for younger pilots flowing into Strategic Air Command's Cold War squadrons.[2]

A Jet Pioneer Before His Peers

Beneath the routine circuits at Bergstrom, Watts carried a pedigree that set him apart from most contemporaries. He was not simply another seasoned piston fighter veteran. In 1945, fresh from Italy and still carrying the dust of war, he had strapped into America's first operational jet fighter, the Lockheed P-80 Shooting Star.

Those early test and familiarization flights at Wright Field and

Muroc introduced him to a world few others yet knew. Turbine engines spooled slowly and demanded timing unlike the instant throttle response of propeller aircraft. Hydraulic systems and electrical instruments required discipline rather than improvisation. At Muroc, he had watched test pilots wrestle with flameouts, engine stalls, and the onset of compressibility at unfamiliar speeds.[3]

That experience made him one of the Air Force's quiet jet pioneers. Most officers his age would not touch jets until the mid-1950s. Watts learned their temperament in 1945, at the frontier.

Back at Bergstrom six years later, he understood that every routine Texan sortie carried weight beyond grading other pilots' landings. It was the cultivation of habits that would carry pilots safely into faster, more unforgiving aircraft. His early P-80 time gave him uncommon credibility, a skill he would deploy repeatedly in the years ahead.

Family Life in Texas

The move to Texas marked a step forward for the family as well. Jeanne gave up her engineering career to take on a new identity as the wife of a senior officer. Their daughter entered elementary school as a military child, absorbing the routines of base housing, commissaries, and a community that measured life in tours and transfers.[2]

Compared to the Charleston and Raleigh trailer years, Bergstrom offered comfort and stability. The family had space, permanence for their daughter, and a sense of belonging within an Air Force that was becoming America's front line.

On 8 September 1951, Watts's Form 5 recorded his instrument rating restored to "White."[3] This clearance authorized him to fly under full instrument conditions, day or night, and made him a key evaluator of younger pilots.

Beyond the ready room, the Korean War still raged. MiG-15s challenged B-29 Superfortresses over the Yalu, and stalemate hardened into terrain. In Texas, America's answer took quieter form: short Texan sorties, instrument checks, and evaluations, a pipeline of precision and reliability built deliberately.

Beneath the Flights

By 1952, Watts's responsibilities expanded. He spent the summer at Chanute Air Force Base, Illinois, where his instrument clearance was further upgraded to "White/Experimental," granting authority not only to fly but to evaluate emerging procedures.[4] This marked his transition from practitioner to instructor, shaping younger aviators as the Air Force evolved from wartime improvisation toward Cold War standardization.

Meanwhile, the 522nd Squadron itself shifted titles with policy debates in Washington: Fighter Squadron, Fighter-Escort Squadron, Strategic Fighter Squadron.[5] Each change reflected uncertainty over whether fighters were tactical tools, bomber escorts, or nuclear delivery systems. For Watts, cockpit routines remained constant: takeoff, navigate, evaluate, land. On paper, however, doctrine rehearsed for a possible third world war.

A Mixed Fleet

Watts's logbooks from 1952 to 1953 show a widening portfolio. Alongside Texans, he flew the veteran C-47 Skytrain and larger C-46 Commando, mastering multi-engine transports while supervising ferry missions and training officers in logistics flying.[6]

Entries also record time in twin-engine trainers such as the T-7 Bobcat and T-11, staples of instrument and multi-engine instruction.[7] His role extended beyond maintaining personal currency. He logged hours as check pilot, evaluator, and instructor, building proficiency across the formation.

Then the jets returned. Republic F-84G Thunderjets and Lockheed T-33 Shooting Stars entered his logbook.[8] [9] For Watts, they represented continuity rather than novelty. He was returning to jet technology first encountered in 1945. That early exposure set him apart. He was now one of the few Cold War flight leaders fluent across piston trainers, transports, and jets alike.

Shifting World, Steady Hand

World events mirrored his evolution. January 1953 brought Eisenhower's inauguration and John Foster Dulles's call for "massive retaliation."[11] In March, Stalin died, shaking the Soviet system built on his authority.[10] In July, the Korean Armistice ended years of stalemate.

For pilots at Bergstrom, such changes appeared mainly in headlines. In his logbook, entries remained steady. Watts understood the connection. Discipline was strategy. Every check ride and midnight Texan sortie fed the larger architecture of deterrence.

By the end of 1953, Watts had become less the combat pilot of Italy and more a professional teacher of aviation. His blend of early jet experience, engineering education, and wartime credibility positioned him uniquely within a Cold War Air Force focused on institutional training and nuclear readiness.

What began as routine circuits in Texans had grown into something larger. He was shaping the flying habits of an entire generation. More than a decade after landing in Casablanca, Watts was once again flying for history. This time, by making others ready.

Notes

1. Blanchard King Watts, *AF Form 5, Individual Flight Records, 42d Air Division, Bergstrom AFB, 1951–53*
2. *BK Watts and Jeanne Johnson Watts 1951 news* (SPOT0002.jpg; SPOT0003.jpg).
3. Watts, *AF Form 5*, 1945–46, P-80 hours at Wright Pat/ MurocWatts, *AF Form 5, Summer 1952 entries, Chanute AFB*, "White/Experimental" upgrade Watts, *AF Form 5 headers, 1952–53 squadron redesignations* Watts, *AF Form 5, 1952–53*, C-47 Skytrain flights.
4. Watts, *AF Form 5, 1952–53*, C-46 Commando, T-7, T-11 flights.
5. Watts, *AF Form 5, 1952–53*, F-84G Thunderjet flights.
6. Watts, *AF Form 5, 1953*, T-33 Shooting Star flights.
7. "Stalin Dies in Moscow at 73 of Cerebral Hemorrhage," *New York Times*, 6 Mar. 1953, 1.
8. John Foster Dulles, "The Evolution of Foreign Policy," *Council on Foreign Relations*, 12 Jan. 1954; NSC 162/2, "Basic National Security Policy," 30 Oct. 1953, *FRUS, 1952–1954*, vol. II.

Chapter 28

Back to China

On 13 May 1953, the 42nd Air Division at Bergstrom issued orders for Lieutenant Colonel Blanchard K. Watts: 180 days' temporary duty with the Military Assistance Advisory Group, MAAG Team MTT 53-28, in Taipei, Formosa, known today as Taiwan. Americans at the time still called it China. This was the United States advisory mission organized to train, equip, and assist allied forces under the Mutual Defense Assistance framework and President Eisenhower's "New Look." His clearance was Top Secret, and the tasking stated plainly: provide on-the-job training for Chinese Air Force maintenance of F-84G aircraft.[13]

This was not a combat deployment but a strategic assignment, part of Eisenhower's effort to build allied capacity rather than commit American forces directly in every conflict. For Watts, it marked something more personal. It was formal recognition of a role that had been taking shape since Bergstrom. His job was no longer to fight the next war, but to teach others how to fight theirs.

What distinguished him among the MAAG advisors was a credential earned years earlier. At Wright Field and Muroc in 1945, he had flown the P-80 Shooting Star, America's first operational jet fighter, before most officers had ever seen a turbine. That experience, buried in old flight logs, gave him an early mastery of jets while the Air Force itself was still deciding what they meant. Taiwan would be the first posting to fully exploit that expertise. He arrived as an advisor. In practice, he was a teacher.

Tainan Air Base: Jets for a New Air Force

By the end of May 1953, Watts reported to Tainan Air Base, where the Republic of China Air Force First Fighter Bomber Wing had been issued Republic F-84G Thunderjets under American aid. The mission was urgent. Defend the island against Communist attack across the Taiwan Strait, barely one hundred miles away.

Watts was appointed to lead the American team charged with training Chinese Nationalist pilots and maintenance crews. The partnership was born of necessity. Mechanics trained on piston engines now faced hydraulics and turbines. Electrical systems failed under load. Gauges were misread. Technical manuals circulated in English that few could read without constant translation. Pilots were confident but uncertain, rigging their first jet flights with a mix of bravado and caution.[2]

Watts taught by doing. He bled hydraulic lines, torqued fittings, and ran instrument sequences himself. Then he stepped aside with the same instruction every time: "Now you do it." He was not there to perform the work. He was there to make others proficient.

Cockpit Proof

Some problems would not yield on the hangar floor. One F-84, notorious for recurring hydraulic failures, had been grounded for weeks and was eroding confidence across the wing. Watts strapped in, ran the checks, and took it airborne. At altitude he rolled, pulled G's, and watched the pressure gauge. This time the needle held.

Back at Tainan he climbed out and gave a crisp thumbs-up. The airframe that had been a liability was now flightworthy. He beckoned a young ROC captain, walked him through the startup, and repeated the instruction that had become his hallmark. "Now you do it." The translator echoed him in Mandarin. The captain launched, flew, and returned safely. It was the first step from hesitation to mastery. That pattern of proof followed by practice defined Watts's method.[2]

Beneath the Flights: Policy and Reality

While Watts trained pilots and crews, policy hardened into structure. The new administration's doctrine emphasized nuclear-backed strength and allied capacity over large standing armies.[3] That shift did not remain abstract for long. It filtered into budgets, basing decisions, aircraft assignments, and training priorities..

To American leadership, Taiwan was a position on a global board. In Tainan, it was a classroom where policy became practice. Strategy was written in the White House. Its success depended on what a young Chinese pilot learned in an F-84 cockpit.

Fig 28-1 Taiwan, 1953. Madame Chiang Kai-shek (with flowers) and Generalissimo Chiang Kai-shek seated at a low central table, surrounded by Chinese staff. One American officer, crouching with his right side to the camera, is visible. (Author's collection)

Colonel Vance's Words

The clearest accolade for Watts's work came from Colonel Reginald C. Vance, MAAG Air Section Chief. His endorsement cut through ceremony. Watts's instruction and supervision were central to "the first conversion of the Chinese Air Force to jet aircraft."[1] U.S. policy marked Taiwan as a bulwark. Chiang Kai-shek's air group needed Americans who could teach as well as fly. Watts delivered both.

Beneath the official accolades, a quieter artifact endured. In the spring of 1953 in Taipei, Chiang Kai-shek and Madame Chiang presided over a formal banquet honoring American air officers who had helped bring the Republic of China Air Force into the jet age. In one surviving photograph from Watts's papers, taken from his own camera position on the left, the Generalissimo and Madame Chiang sit at a low central table, framed by a dense arc of Chinese staff. Only one American officer is visible, kneeling with his back to the camera, likely presenting the flowers Madame Chiang holds. The angle is candid rather than staged, as if Watts were trying to capture a real moment rather than a headline

On Watts's wrist that afternoon, barely visible at the cuff, was a gold Rolex. In later years it wore a plain Twist-O-Flex band, the mark of a workman's tool rather than a showpiece. Family memory holds that it was not purchased but presented, a personal gift from Generalissimo Chiang Kai-shek to the American who had given the Chinese Air Force a way forward.[2] No official corroboration survives. The watch remains, its face carrying a soft patina of service, an heirloom that keeps not just time but a story.[6]

Chiang Kai-shek's Long Fade

When BK stood off to the side in Taipei in 1953, watching Generalissimo Chiang Kai-shek and Madame Chiang at the banquet from the wings, the Nationalists' story still presented itself as one of exile and return. Taiwan was the redoubt from which, in official rhetoric, the Republic of China would someday "recover the mainland." The jet pilots Watts helped train at Tainan Air Base were part of that promise. Their F-84 Thunderjets and later fighters were both shield and symbol, proof that Chiang's air arm could modernize, stand beside American allies, and hold the line against the People's Republic across the Strait.[12]

History bent another way.

Through the 1950s and 1960s, Chiang's rule hardened into a

210

one-party state under martial law. The White Terror years brought arrests, imprisonment, and executions of real and suspected opponents.[8] The Republic of China retained its United Nations seat with American backing, but the dream of return receded as the People's Republic consolidated power on the mainland.[9]

By the early 1970s, the diplomatic frame in Taiwan that had defined Watts's tour had collapsed. In 1971, the United Nations transferred China's seat from Taipei to Beijing.[10] The following year, President Nixon traveled to the People's Republic, ending the long-standing fiction of exclusive recognition. When Chiang Kai-shek died in 1975, he was still President of the Republic of China on Taiwan, but the world had moved on.[11]

For the men Watts trained, the arc was different. They did not recover the mainland. Instead, they formed the backbone of what became the modern Republic of China Air Force, defending Taiwan through successive crises and into the missile standoffs of the late Cold War. Their professionalism, shaped in part by American advisors like Watts, outlived Chiang's political ambitions.[12]

In BK's papers, the photographs and the letter from General S. M. Wang awarding Chinese Air Force pilot's wings preserve an earlier moment, when Chiang's image, Madame Chiang's poise, and the jet age at Tainan still seemed to point toward a single, unbroken future.[14] History did not honor that vision. The work itself endured. Pilots were trained. Airplanes flew and an island survived

---

Lessons Carry Forward

By early 1954, AF Form 5 entries show Watts back in Texas, flying the T-33 Shooting Star at Bergstrom, then reassigned to the 516th Strategic Fighter Squadron at Great Falls (now Malmstrom) AFB, Montana. Through spring and summer the logbook filled with C-47 Skytrain ferry flights and check rides. It looked routine, but it was the continuation of the role crystallized in Taiwan.[5]

At the beginning of his career, Watts had been a combat pilot. At Bergstrom he became a professional evaluator. At Tainan he stepped fully into his postwar identity, the man who taught others to master the jet age. In Great Falls, that teacher's role would expand into the doctrinal center of Strategic Air Command.

Notes

1. Headquarters 42d Air Division, "Orders," LO 348, 13 May 1953, Bergstrom AFB, TDY Taipei, MAAG Team MTT 53-28 This document details Watts' temporary duty assignment to Taiwan.

2. Headquarters CAF 1st Fighter Bomber Wing, "Retention of Lt. Col. Watts," 13 August 1953; Headquarters CAF (Gen. S. M. Wang), "CAF Wings No. 72047," 7 September 1953; CAF Wing, "Gift of Appreciation," 4 December 1953. These documents, found in the Author's Collection, commend Lt. Col. Blanchard K. Watts's "outstanding instruction and supervision" in the introduction of F-84G Thunderjet operations and credit his work as central to "the first conversion of the Chinese Air Force to jet aircraft."

3. National Security Council, "Basic National Security Policy (NSC 162/2)," 30 October 1953, Foreign Relations of the United States, 1952–1954, vol. II; John Foster Dulles, "The Evolution of Foreign Policy," Department of State Bulletin 30 (25 January 1954): 107–10. For the broader geopolitical context of the period.

4. "Agreement at Panmunjom: Armistice Signed July 27, 1953," New York Times, 28 July 1953; "General John E. Hull to Succeed Mark Clark in Far East Command," New York Times, 31 July 1953; "Arthur Radford Named Chairman Joint Chiefs," New York Times, 16 August 1953. These articles provide a snapshot of key events and appointments during Watts' time in Taiwan.

5. Blanchard King Watts, AF Form 5, Individual Flight Records, February–July 1954, T-33A entries; reassignment to 516th Strategic Fighter Squadron (March 1954); C-47 Skytrain flights

(April–June 1954) These records detail Watts' flight activities and assignments following his Taiwan tour.

6. Family oral history, Watts family papers, private collection. While the gift of the Rolex from Generalissimo Chiang Kai-shek is a cherished family legend, no corroborating evidence has been found in official U.S. or ROC records to date.

7. Jay Taylor, The Generalissimo. Chiang Kai-shek and the Struggle for Modern China (Cambridge, MA: Harvard University Press, 2009), esp. 353–381. For Chiang's post-1949 exile narrative and rhetoric of "recovering the mainland," and for the symbolic role of the air force in early Taiwan.

8. George H. Kerr, Formosa Betrayed (Boston: Houghton Mifflin, 1965), 247–310; Fredrik Fällman, "The White Terror in Taiwan," in The Routledge Handbook of the Cold War in East Asia, ed. Zheng Yangwen and Hong Liu (London: Routledge, 2018), 191–208. These sources provide context for the "White Terror" period in Taiwan.

9. Taylor, The Generalissimo, 395–432; Nancy Bernkopf Tucker, Strait Talk. United States–Taiwan Relations and the Crisis with China (Cambridge, MA: Harvard University Press, 2009), 16–45. For the ROC's UN position and the fading prospect of "return."

10. Margaret MacMillan, Nixon and Mao. The Week That Changed the World (New York: Random House, 2007), 181–239; United Nations General Assembly Resolution 2758 (XXVI), 25 October 1971. These sources detail the shift in international recognition of China.

11. Taylor, The Generalissimo, 433–452; Shelley Rigger, Why Taiwan Matters. Small Island, Global Powerhouse (Lanham, MD: Rowman & Littlefield, 2011), 41–67. For the later political developments in Taiwan.

12. Kenneth W. Allen and Kenneth F. W. Allen, Taiwan's Defense Reform and the Chinese Air Force (Washington, DC: RAND

Corporation, 2007), 9–23. For the development of the modern Republic of China Air Force.

13. Steven M. Goldstein and Julian Chang, eds., Taiwan's Presidential Politics. Democratization and Cross-Strait Relations in the Twenty-First Century (Cambridge, MA: Harvard University Asia Center, 2008); Rigger, Why Taiwan Matters, 121–166. For Taiwan's democratization.

14. Blanchard K. Watts papers, family collection. Citation to "Letter from General S. M. Wang to Capt. Blanchard K. Watts, awarding Chinese Air Force pilot's wings," 1953, and associated photographs from Tainan Air Base.

CHAPTER 29

STRATEGIC CURRENTS

By early 1954, the ink on BK's Form 5 shifted headings once again. His brief interlude at Bergstrom Air Force Base after his second China tour allowed him to re-ground stateside, recalibrating with turbine flights and the cadence of Strategic Air Command paperwork. Soon, another order arrived. The logbook header changed to Great Falls Air Force Base, Montana, and the 516th Strategic Fighter Squadron.[1]

It was a lonely posting on the fringe of America's interior. Winds swept unbroken plains. Winter skies flattened into gray sheets that held for weeks, lightless but charged with static. For Watts, the austerity was not exile. A farm boy from North Carolina understood empty fields and weather rhythms. Montana demanded endurance, but it also offered familiarity. Vastness felt less alien than he had remembered of a childhood transposed into wide sky.

A Commendation Hand-Delivered

February brought recognition from halfway around the world. A formal commendation from the Chinese Air Force, delayed for months in transit, arrived in Montana. In it, General S. M. Wang recorded that Lieutenant Colonel Blanchard K. Watts had rendered valuable assistance in jet fighter training at Tainan and directed that he be presented a pair of Chinese Air Force pilot's wings, serial number 72047.

215

(TRANSLATION)

HEADQUARTERS CHINESE AIR FORCE

TAIPEI, TAIWAN, CHINA

C
O
P
Y

7 September 1953

SUBJECT   : The Presentation of a CAF Pilot's Wings.

TO   : Lt. Col Blanchard K. Watts

THRU   : Colonel Edward F. Rector
             Chief, Air Section, MAAG

FROM   : General S. M. Wang
             Commanding General, CAF

According to a report submitted by Colonel Liu Wei-cheng, Commander of the 1st Wing stationed at Tainan that Lt. Col Blanchard K. Watts in the course of his service as an advisor at Tainan AFB has rendered valuable assistance in jet-fighter training.

This Headquarters takes the pleasure to express its gratitude to Lt. Col Blanchard K. Watts for his outstanding assistance and to present to him a pair of the Chinese Air Force Pilot's Wings (Serial No. 72047) with the hope that opportunities will permit this Headquarters to avail ourselves of his further assistance in the future.

S. M. WANG
General, CAF
Commanding General

Fig. 29–1. Delivered to BK in the spring of 1954, this letter from General S. M. Wang, Commanding General of the CAF, commends Watts for his "valuable assistance in jet-fighter training" at Tainan and awards him a pair of Chinese Air Force Pilot's Wings (Serial No. 72047). (Author's collection)

Colonel Einar A. Malmstrom, the decorated deputy commander of the base and a P-47 veteran of Europe, walked the letter into Watts's hands and into his personnel file.[3]

The exchange was understated, almost routine, yet it carried weight. A World War II combat leader whose record was written in flak and thunder placed an accolade for patience, teaching, and precision into the hands of a pilot who had carried those virtues across Taiwan's hot tarmacs. The Air Force itself was changing, from explosive combat toward endurance, training, and deterrence. Within a year, Malmstrom would be lost in an F-86 crash and the base would bear his name. For Watts, the handoff became a quiet bridge between eras.

Col. Einar Axel Malmstrom (1907–1954)

- Wisconsin native; P-47 combat pilot, 356th Fighter Group.
- Decorations: Distinguished Flying Cross, Air Medal.
- Deputy Commander, Great Falls AFB; personally delivered Watts's Chinese AF commendation, Feb. 1954.[3]
- Killed 21 Aug. 1954 in F-86 crash. Great Falls, AFB renamed Malmstrom AFB in his honor, Oct. 1955.[12]
- Significance: A wartime giant; his presence cast Cold War professionals like Watts in sharper light.

Giants of the Era

Strategic Air Command loomed with names that defined the Cold War.

General Curtis E. LeMay, architect of American strategic bombing in World War II and commander of SAC since 1948, imposed readiness as culture. His inspections were feared for their severity. A single missed item could ground crews. He demanded readiness measured not in theory but in minutes, with teams sleeping beside aircraft, bombers fueled and poised against Arctic winds.

Above LeMay in Washington stood General Nathan F. Twining, Air Force Chief of Staff and later Chairman of the Joint Chiefs. Twining translated doctrine into global posture, bending flight arcs over the pole with rigid logic. He turned nuclear strike planning from contingency into doctrine, making SAC the backbone of containment.[1]

Then came General Thomas S. Power, LeMay's protégé and

successor, whose vision pressed deterrence toward overwhelming first strike. Twining drew the outlines. Power hardened them.

Measured against these figures, Watts stood quietly in the margins. He filled no headlines. His labor was execution: ferrying Thunderjets through Montana weather, maintaining proficiency in transports and trainers, logging flight records in neat script. If Washington sketched policy and LeMay dictated culture, pilots like Watts made deterrence real. His seat in the cockpit, his shoes on the hangar floor, his signature on the Form 5 were the margins where strategy became daily work.

The logbook shows a mosaic of airframes.[2] Watts ferried C-47 Skytrains and C-46 Commandos through snow squalls and iced prairie. He kept turbine instincts sharp in the T-33 Shooting Star, linking back to early P-80 flights at Wright Field and Muroc in 1945.[3] He returned to the F-84G Thunderjet, now harnessed by SAC for tactical nuclear delivery.[3]

Together the types reveal not breadth alone, but formation. His versatility hardened into discipline.

Frozen Dawn

One February morning, hangar doors stood rimed with ice, rollers grinding against the cold. Watts signed the order, climbed into a C-47, and coaxed life from its engines. The heater failed. Wheels clawed through slush, spraying arcs that froze before soaking the blacktop as the transport lifted into a gray horizon.

At altitude, the steady hum of Pratt & Whitney engines filled the cockpit, familiar since cadet days. Inside the Dakota, Watts adjusted trim and fought a gust that rocked the fuselage. The mission was plain: deliver supplies eastward. Routine on paper, yet no sortie under SAC was truly routine. In deterrence, even ferry flights threaded into strategy like pins in a cold geometry mapped in Omaha.

Only later, on the radios and in the papers folded into flight bags, did the larger world intrude. On 1 March, Castle Bravo detonated at Bikini Atoll, its yield far beyond forecast, fallout drifting across the Pacific.[4] Senator McCarthy's campaign unraveled under televised inquiry. Not long after came darker dispatches:

Dien Bien Phu collapsing under Viet Minh fire, the French tricolor falling, demonstrating France's Indochina empire breaking.[5] From Geneva came lines on paper dividing Vietnam at the seventeenth parallel.[6]

Hunting the High Plains

Beyond the cockpit, other memories formed. For Watts's eldest, Montana meant sharp air, elk and antelope hunts, and snow crunching beneath boots. They camped in tents sewn from parachute silk, rifles propped outside: his bolt-action .30-06, her lever-action .30-30, both still held by the family.

For a teenage girl, the romance came later. In the moment there was cold that rose from the ground before dawn and settled in bone. Parachute-cloth walls snapped in the wind, thinner than a bedsheet. Even with two pairs of socks, toes ached. They woke to frost edging the zipper and ice on the canteen.

What made it bearable was that it was theirs, father and daughter against weather. In the dark he would murmur, "You warm enough, baby girl?" then he would start the fire with practiced economy. By the time she crawled into the bitter air, coffee hissed in a tin pot and warmth bled from coals into snow.

On the icy runways, Watts was the reserved professional among SAC's titans. In timberline clearings, however, he was the father who taught stillness before a shot and the quiet craft of making fire in the snow. Montana held both: frozen runways tethered to Moscow on command charts, and quiet plains where endurance was learned one cold night at a time.

Transition Toward Teaching

By 1956 the record shifts again. Watts formally assumed instructor responsibilities.[8] Vance's praise in Taipei and Malmstrom's handoff in Montana anticipated the turn. "Teacher" became designation as much as it was also his nature.

Montana's skies were gray, but beneath them stretched both SAC deterrence and family laughter around campfires. Here, Watts walked with a different set of giants, anchored in the same duty: doing the work and ensuring his planes could fly at dawn.

Notes

1. Blanchard King Watts, AF Form 5, Individual Flight Records, 1954: transfer to 516th Strategic Fighter Squadron, Great Falls AF).

2. Watts, AF Form 5, 1954–56 entries: C-47 Skytrain and C-46 Commando sorties.

3. Headquarters Chinese Air Force, General S. M. Wang, "The Presentation of a CAF Pilot's Wings," 7 September 1953 (translation), Taipei, Taiwan, to Lt. Col. Blanchard K. Watts, through Colonel Edward F. Rector, Chief, Air Section, MAAG. The letter cites Watts's "valuable assistance in jet-fighter training" at Tainan and directs that he be presented "a pair of the Chinese Air Force Pilot's Wings (Serial No. 72047).

4. "U.S. Atomic Energy Commission, Castle Bravo Report, Bikini Atoll, 1 Mar. 1954.

5. Bernard Fall, *Hell in a Very Small Place: The Siege of Dien Bien Phu* (NY: Da Capo, 1966).

6. "Final Declaration of the Geneva Conference," 21 Jul. 1954.

7. John Lewis Gaddis, *Strategies of Containment* (Oxford: OUP, 1982), chap. 5.

8. AF Form 5, May–June 1956, instructor certification.

9. Curtis E. LeMay with MacKinlay Kantor, *Mission with LeMay: My Story* (Garden City: Doubleday, 1965).

10. Nathan F. Twining, *Neither Liberty Nor Safety: A Hard Look at U.S. Military Policy and Strategy* (New York: Holt, Rinehart and Winston, 1966).

11. Thomas S. Power, *Design for Survival* (New York: Harper & Row, 1965).

12. U.S. Air Force Historical Research Agency, biographical file: Col. Einar A. Malmstrom (1907–1954).

The next decisive turn in BK's career came in May 1956. On AF Form 5, struck in the clean letters of a base typewriter, a single line reframed his career. Instructor Pilot, C-47 Skytrain.[1] From that moment the entries changed tempo. The logbook ceased to be a ledger of solo hours and became the score of a teacher logging dual flights and check rides, evaluations recorded in columns that decided whether a young pilot would steady his hands at night or chase the runway lights into panic.

Across three years, more than 436 of those hours stacked up over Montana and the northern routes. Crosswind landings crabbed the C-47, the "Gooney Bird," sideways into scouring gusts. Night instruments reduced the world to needles and trust. Then came engine-out drills where the left prop feathered into stillness and the right pulled like a plow horse, demanding rudder and nerve.[1] "Now you do it," he said, and the airplane, ungainly in profile yet loyal in service, taught the rest.

The work was quiet. It was consequential. SAC's bombers required fuel, spares, and motion. The C-47s and their successors carried that lifeblood, and BK became a teacher who forged the pilots that kept the arteries open in any weather.

Family Expands on the Plains

Montana pressed its own pattern onto the household. Two more children were born in Great Falls, adding to a home already full of small coats on pegs and a kitchen chalkboard dense with base schedules. Jeanne ran the interior campaign. Doctor visits in snow. BX runs. The art of coaxing heat from coal and patience from toddlers while her husband graded students and briefed check rides.

Routine defined the days. Buses shouldering through snowdrifts. Brown paper packages stamped with APO numbers. Saturday mornings devoted to boots, oil stains, and the smell of canvas. Yet even everyday routines in SAC lived under a conditional mood: the readiness drill, a klaxon that might test a squadron's promise that it could move at once. The family laughed, learned, and slept beneath the potential for air raid sirens at any moment of the day or night.

For BK's eldest, the tension seeped in at child height. At school in Great Falls, she took part in air-raid drills where classes filed down into bomb shelters, practicing for a flash and shock wave no one could quite explain without scaring the children. The concrete corridors smelled of dust and disinfectant, and the teacher's voice stayed bright and brisk as they counted off and sat cross-legged on the floor.

Into the Regular Air Force

In April 1958, permanence arrived on paper. A Department of the Air Force order appointed him a Lieutenant Colonel in the Regular Air Force.[2] He was no longer a wartime reservist recalled for utility but a career officer entrusted with continuity and command.

Less than a year later, the silver eagles of Colonel arrived, along with orders to Air University, Maxwell AFB, Montgomery, Alabama. Air War College.[3] The instructor who could settle a shuddering transport at night would now be asked to debate the structure of nuclear peace.

Cold Geometry

One midnight shortly before he left Great Falls, he found himself over the Hi-Line, cloud layered black upon black. The Gooney Bird's cockpit glowed a dim green. The heater grudged against the cold. To the south, a radio snip, arguments over budgets and missiles. To the west, another snip, a test shot on Christmas Island postponed for weather. To the east, a murmur about Cuba's revolution tightening its grip.

The airplane did not care. Needles, headings, altitudes were its language. His student drifted off-course. Watts reached up and rapped the compass glass with a knuckle, the same old trick Ben Moore had used on balky gauges at Hicks back in '41. "Airplanes talk" if you listen. BK said nothing. After a beat, the young pilot corrected.

"Now you do it" was not a slogan. It was the quiet confidence that a system would work if its people did.

The Teacher Becomes the Student

Maxwell AFB reset the frame. In early spring 1959, Watts took a seat beneath the portraits of General Henry "Hap" Arnold and General Carl A. Spaatz, the general who handed him his first Air medal back in North Africa. The sense of coming full circle was surely with BK that afternoon in Montgomery. The Air War College, founded in 1946, had matured in the decade of brinkmanship into the Air Force's top school for senior professional education.

Classrooms rang with the vocabulary of the era. Massive retaliation. Counterforce. Flexible response. Alliance entanglements and escalation ladders. Bomber penetration versus missile survivability. Men who had flown P-51s over Europe and F-86s over Korea now debated fallout patterns and second-strike credibility.[5] The ethos was stern and abstract, but it was built on familiar bones. Discipline. Redundancy. Timing.

Montgomery itself made the abstraction urgent. The Bus Boycott of 1955–56 had already launched a young pastor, Martin Luther King Jr., from Dexter Avenue Baptist Church into history. By the time the Watts family arrived, sit-ins and boycotts had multiplied across the South. Freedom Riders would soon wheel into the bus terminals. Federal courts wrestled with de facto and de jure while state power flexed to resist.[12]

To drive out of base housing in the afternoon was to cross a seam in American life, staff cars were easing past placards and patrols. Inside the seminar rooms, officers parsed how to prevent nuclear war. Outside, citizens demanded the nation fulfill its own promises. The contradiction did not resolve itself, it simply framed every day for them.

Recognition in Montgomery

Amid doctrine and discord, recognition found him again. In March 1960 the Air Force approved Colonel S. P. La Barrera's recommendation awarding Watts the Air Force Commendation Medal.[4]

The citation reached backward and outward at once. Reorganizing the 4061st Air Refueling Wing's maintenance structure at Malmstrom. Training scores of younger pilots to proficiency. Knitting together depots, supply, and squadrons so that conversions to new SAC models did not tear seams. The medal was presented where he now studied strategy in Montgomery, fusing his two halves, the teacher of machines and men and the commander of systems.

Toward the Heavy Years

By mid-1960, graduation orders carried him forward. The next billets would be heavy metal and heavy consequence. Biggs AFB in Texas and Castle AFB in California, where the worlds of the B-52 Stratofortress and the elaborate choreography of nuclear readiness intertwined.

The years of flying the C-47 had not been an interlude. They had been formation. The making of a colonel who could compress the intimate patience of instruction into the brutal timing of deterrence.

He left Maxwell AFB with the same quiet posture he had carried into it. No headlines. No speeches. Just the sense that dawn would come, checklists would be read, systems would align, and somewhere an airplane would lift exactly when it should.

Notes

1. Blanchard King Watts, AF Form 5, entries 1956–59, annotated "Instructor Pilot," >436 instructor hours.
2. Department of the Air Force, Personnel Orders, April 1958, appointment to Regular Air Force.
3. Department of the Air Force, Orders, January 1959, promotion to Colonel and assignment to Maxwell AFB / 3800th AB Wing.
4. Department of the Air Force, "Air Force Commendation Medal Citation, approved March 1960".
5. For the evolution of Air War College curriculum in the late 1950s and early 1960s and its focus on nuclear strategy, deterrence, and the shift from "massive retaliation" to "flexible response," see Walter S. Poole, Adapting to Flexible Response, 1960–1968 (Washington, DC: Office of the Secretary of Defense, 2013), esp. chap. 1; and John B. Wilson, "Maneuver and Firepower: The Evolution of Divisions and Separate Brigades," CMH Pub 60-14 (Washington, DC: U.S. Army Center of Military History, 1998), chap. 12, "Flexible Response." Both trace how professional military education incorporated nuclear strategy concepts that were reflected in the Air War College syllabi BK encountered when he enrolled in early spring 1959. Poole PDF via DTIC; "Flexible response" overview.
6. U.S. Department of State, "The Suez Crisis, 1956," Office of the Historian.
7. U.S. Department of State, "The Hungarian Revolution of 1956," Office of the Historian.
8. National Security Council, "Report to the President by the Security Resources Panel of the Science Advisory Committee (Gaither Report)," 7 November 1957, declassified and edited version, in Foreign Relations of the United States, 1955–1957, vol. XIX, National Security Policy; Information Policy; Arms Control and Disarmament (Washington, DC: Government Printing Office).

9. NASA History Division, "Sputnik and the Dawn of the Space Age," October 1957.

10. U.S. Department of State, "The Eisenhower Doctrine and the 1958 Lebanon Crisis," Office of the Historian.

11. Office of the Secretary of Defense, "U.S.–Taiwan Strait Crises: 1954–55 and 1958," historical brief.

12. U.S. Army Center of Military History, Advisory Efforts in Vietnam, 1955–60 (CMH publications, MAAG-Vietnam summaries).

13. Federal Judicial Center, "Civil Rights and Federal Courts: Montgomery and the Bus Boycott, 1955–56"; Library of Congress, Civil Rights History Project (context for 1959–60 sit-ins and Freedom Rides in Montgomery).

Chapter 31

The Last Cockpit

Darkness pressed against the canopy. It seemed endless and absolute until the only world remaining was the cockpit's dim cathedral of light: radar scopes breathing green, attitude bars steady as winter stars, needles settling like snow. Beyond the pressure hull, the Arctic rolled away unseen, a frozen continent scrolling like a map beneath their feet. Eight turbojets droned in a braided hum that found its way through bone and helmet, the basso continuo of Strategic Air Command.

Colonel Blanchard K. Watts sat harnessed into the pilot's chair in the cockpit of a B-52D Stratofortress, one of SAC's heavy bombers sometimes known as its "Sunday Punch."[1] This was a machine built to carry the end of things. Intercoms murmured. Checklists droned. Fatigue layered hour upon hour. A tanker rendezvous, a polar fix, a bomb-release check, gunnery-station tests and pressure-system scans punctuated the long monotone. Routine, they called it. But routine in a B-52 meant rehearsing the unthinkable until it became muscle memory.

Two decades earlier, he had flown open-cockpit biplanes, wearing a leather helmet pulled low against slipstream and cold. Now he sat in a sealed, pressurized hull larger than some houses, flying a weapon that could redraw maps. In that arc, from wood and fabric to nuclear steel, you could read the Air Force's own passage of time and its technological advance.

Fig 30-1 Photo: 1960 B52 (likely variant D) on the ground at Biggs AFB. (Author's collection)

Transfer to Biggs

The desert expanse of Biggs Air Force Base, El Paso, Texas, was a SAC stronghold with concrete poured long and hard. Runways stretched more than two miles, and the horizon became a straightedge in every direction.[1] Biggs had ushered in the B-36's age and now tuned itself to the B-52's longer reach, where global operations were planned in rooms where clocks ran in Zulu and maps curved across the pole.

Before Biggs would trust him with a Stratofortress, the syllabus ran through California at Castle AFB, the forge where heavy crews were made.[2] Transition was precise, punishing, and without sentiment. Polar navigation in a world without landmarks. Emergency drills scripted to the second. Tanker rendezvous in weather that swallowed distance. ECM and crew coordination drilled until twelve hands felt like one. By late 1960, he was back at Biggs with the qualification in his jacket and the big jet's rhythm in his ear.[2]

The Boeing B-52 Stratofortress

- First flight: 1952; operational 1955
- Engines: 8 Pratt & Whitney turbojets
- Wingspan: 185 feet
- Crew: 6–8 officers and enlisted
- Range: 8,800+ miles unrefueled; global with KC-135 tanker support
- Payload: up to 70,000 lbs, including nuclear weapons
- Cold War role: SAC's intercontinental "Sunday Punch," rehearsing polar strike profiles without cease

Life in the Stratofortress

Everything about the B-52 redefined scale. Where C-47 instruction had counted hours in ones and twos, the Stratofortress ran in tranches of six, eight, even ten hours logged on a routine day.[3] Takeoff alone was an orchestration: eight engines spooling, brakes braced, then released as iron surged forward until the wings, impossibly long, lifted and flexed skyward, as though remembering a bird's first instinct.

The routes bent north into thin air. Polar rehearsal was not theatrical. It was geometry and stamina. The shortest arc from Texas to targets in the Soviet Union crossed the top of the world, so SAC practiced that line until crews and aircraft could draw it blind.[3] The cockpit lived between tedium and terror: hours of steady-state engine hum, then bursts that demanded total attention to timed switches, crisp calls, and hands moving in small, exact choreography across panels and throttles.

Endurance itself had become a weapon. Deterrence demanded proof. Not abstract promises, but crews who could stay sharp into the tenth hour, whose checklists matched doctrine, whose tanker hookups ghosted into place at the scheduled minute. The weapon was credibility. The delivery system was that fatigue resisted and was overcome.

Night over the Arctic could feel like flying inside a bell. A tanker rendezvous brought a second ship's lights into their small universe, two constellations drawing into alignment in a place without horizon. When the refueling boom latched, vapor arced. A shudder ran down the fuselage as fuel crossed the narrow distance between two nations' promises. The navigator called time. The pilot barely moved the yoke. The crew breathed in unison until disconnect. This was the sinew of strategy, not the headline. This dance repeated itself miles above the middle of nowhere, along polar routes only giants were asked to tread.

Cold War Shadows

Politics below thrummed against the canopy's glass, distant but unmistakable.

- 1 May 1960: Gary Powers's U-2 fell from the sky over Sverdlovsk. Eisenhower's Paris summit collapsed beneath the wreckage and a captured pilot; the fiction of American invulnerability died at altitude.[4]
- Autumn 1960: In New York, Nikita Khrushchev's shoe pounded a UN desk and the world listened to theater as strategy, signal wrapped in spectacle.[5]
- November 1960: A narrow election placed John F. Kennedy at the helm of SAC's arsenal. New doctrine flirted with "flexible response," but until missiles could guarantee second strike, the bomber force remained the visible fist.
- 1961: Berlin boiled; by August, concrete and wire cut a city in half. The Wall rose, and every alert on every base felt less abstract.

Toward Command Pilot

BK's AF Form 5 entries for 1960 and 1961 read sparely, almost indifferently: date, tail number, crew initials, block hours and equipment.[3] But the clusters told the story. Hours stacked in the

Stratofortress. Endurance missions flown along polar lines. A senior pilot who could settle a large, long airplane into the sky with calm and bring it home on the numbers.

His handwritten log goes quiet after 1961. The record does not.

On 13 January 1962, the Air Force conferred the aeronautical rating of Command Pilot on Colonel Blanchard K. Watts, the service's highest pilot qualification.[6] The badge did not arrive as a gift but as an audit: thousands of hours across multiple types, combat and instruction, transport and jet, responsibility proven through years of evaluation. In public, it meant what the metal wings implied. This man is one of ours at the highest level of the craft. In private, it closed a circle begun in a wooden seat over the prairies of East Texas.

The Last Cockpit

If a life can be read in machines, BK's spanned the century: open biplanes, P-40s over North Africa, long ferries across China, early jets at Wright Field and Muroc, instruction flights in Montana, and finally the iron lungs of deterrence. He had flown from the smallest to the biggest, but the scale was secondary. What mattered was the craft carried intact from one cockpit to the next: calm in the wind, discipline in procedure, and the ability to teach what could not be written.

By mid-1962, new orders arrived. United States European Command in Paris called, drawing him from cockpits into command rooms, from flight lines to alliance tables. Flying was complete. The strategic years had begun.

He closed the log. The B-52 lifted again at dawn, another crew rehearsing the impossible until it felt ordinary. In those final entries, Watts carried an era across its last threshold, from canvas to nuclear steel, from trainee to Command Pilot, and from the intimacy of flight to the cold geometry of power.

Notes

1. Blanchard K. Watts, AF Form 5, assignment record, May 1960: transfer to Biggs AFB.
2. AF Form 5, 1960 entries: Castle AFB heavy transition sylla-

bus and qualification.

3. AF Form 5, 1960–61 entries: B-52 sorties, 6–10 hour endurance profiles, polar routes.
4. "U-2 Plane Shot Down Over Soviet Union; Captured American Pilot Alive," New York Times, 2 May 1960.
5. "Khrushchev Raps West at the U.N.," New York Times, 13 Oct. 1960; shoe-pounding widely reported in contemporary press.
6. Dept. of the Air Force, Orders, 13 Jan. 1962, Aeronautical Rating: Command Pilot.

CHAPTER 32

FOREIGN SERVICE

The journey east over the Atlantic was mythic in its scale. In the summer of 1962, the Watts family boarded the RMS Queen Elizabeth at New York, a floating city under stack steam, bound for France.[1] Dining rooms gleamed with chandeliers. Ballrooms echoed with orchestra music. Promenades ran along the Atlantic horizon line, seemingly without end.

Jeanne guided four children across the decks, her cadence practiced from a decade of duty stations. Their oldest daughter, now approaching college age, looked ahead to studies in Paris. Colonel Blanchard K. Watts, silver eagles still fresh on his shoulders, stood at the rail and turned eastward. For twenty years his eyes had scanned horizons from cockpits. Now the world awaited him across the sea, not as a pilot, but as a senior officer drawn into the Cold War's staff machinery, working within the orbit of United States European Command (EUCOM).

The French Connection

Paris in 1962 sat at the fulcrum of NATO's uneasy balance. It was the headquarters of American and Allied staff, but it was also Charles de Gaulle's Paris, restless and openly skeptical of American dominance. Watts entered the Palais de l'OTAN carrying the same traits that had defined him in cockpits: clarity, candid precision, and a dislike for ornament.

Staff officers noticed quickly. French counterparts, masters of ceremony and form, found his blunt efficiency unexpectedly useful.[2] Letters from that first year praised his ability to smooth Franco-American friction. General officers remarked that where others obscured, Watts translated complexity into action. His command presence, sharpened by years of cockpit authority, transferred cleanly into briefing rooms and committee chambers.

Commendation: Paris, 1962

"Colonel Blanchard K. Watts has contributed exceptional ability and tact in the conduct of his assignment here in Paris. His devotion to cooperation between our forces has been evident and most welcome."
- General Bondley, USAF
  EUCOM Senior Staff
- Lt. Gen. James M. Gavin, USA
  Commander, VII Corps
  The "Jumping General" of World War II

The Cold War's Edge

Soon came Watts's dispatch to Germany, where the United States deployed its new Pershing intermediate-range ballistic missiles.[3] Their presence drew Moscow's fury and Germany's caution in equal measure. The assignment demanded diplomacy as much as tactics. American readiness had to be preserved while German sovereignty was visibly respected.

Watts became the balancing hand. He met Bundeswehr officers with the same discipline he had once taught cadets, pairing formal respect with operational credibility. Commendations followed, bearing the names of men who shaped the Cold War itself.

Commendation: Pershing Missile Support, 1962

"...Colonel Watts distinguished himself by flawless cooperation with American and German missile units. His energy provided Pershing forces with readiness essential to our common defense."
- General Bondley, USAF
  EUCOM Command
- Lt. Gen. G. B. Russell, USA

Deputy Commander, U.S. Army Europe

- General Earle G. Wheeler, USA
  Chief of Staff, U.S. Army
- Generalmajor Kuntzler
  Bundeswehr Missile Command

Iran, 1962: Commendation

Cold War calculation gave way abruptly to catastrophe. In July 1962, a massive earthquake shattered Qazvin, Iran, killing thousands and leaving survivors scattered across a broken plateau.[4] Within hours, EUCOM repurposed its airlift capacity. From Paris came the order: move relief now.

Watts bridged staff planning and execution. C-130s loaded with medicine, tents, rations, and water purification gear thundered east across Turkey and onto the Iranian plateau. On flight lines and in embassy meetings alike, his steady hand aligned priorities. The contrast was stark. One month shaping nuclear readiness, the next shepherding food and shelter to families under collapsed walls."For diligence and efficiency in ensuring relief movements into the affected area, the undersigned conveys appreciation. Colonel Watts's contribution was a measurable factor in the success of American assistance in Iran."

- General Bondley, USAF, EUCOM Commander
- Admiral John C. Hayden, USN; Allied Naval Forces, Southern Europe
- Robert L. Campbell; State Department, Office of Foreign Disaster Relief

The Balkans

The earth shifted again in July 1963. Skopje, Yugoslavia, convulsed under an earthquake that killed more than a thousand and destroyed most of the city.[5] A year later, floods swept the Sava River and submerged Zagreb. In the Balkans, American aircraft carried political weight as well as cargo. Every movement risked ripples in Belgrade's carefully guarded neutrality.

Once again, Watts served as hinge and interpreter. At Skopje he translated urgency into tact, organizing convoys and scheduling C-130 drops that bypassed bureaucratic choke points. At Zagreb he ensured aid moved swiftly without triggering Cold War trip-wires. Diplomats noted his balance. Aid delivered too loudly could become propaganda. Aid delivered quietly could save lives.

Commendation: Yugoslavia, 1964

"…Colonel Watts has been instrumental in organizing humanitarian relief during the earthquake in Skopje and subsequent flooding in Zagreb. His efficiency and tact ensured assistance was delivered swiftly and with sensitivity. The Ambassador conveys his sincere thanks."

- Frank Rouse, Deputy Chief of Mission, U.S. Embassy Belgrade
- Gen. Jacob E. Smart, USAF, Commander U.S. Air Forces Europe (later Vice Chief of Staff, USAF)
- C. Burke Elbrick, U.S. Ambassador to Yugoslavia

Back Toward Teaching

The arc closed in June 1965, not with ceremony for its own sake, but with recognition from peers who understood the work. At Camp des Loges, outside Saint-Germain-en-Laye, Watts received a Joint Service Commendation Medal, presented on 4 June 1965 by Captain Robert W. Schepers, USN. The two men had become friends through shared assignments and shared assumptions about service.

Competence mattered more than visibility, and continuity mattered more than credit. The award marked the end of Watts's extended foreign service, closing a career phase defined by coalition work, technical judgment, and trust across uniforms. With that chapter complete, he turned homeward, returning not to retirement but to instruction, carrying the accumulated weight of operational experience back into the classroom at Rensselaer Polytechnic Institute.

Fig 32-1 Captain Robert W. Schepers, USN showing off the Joint Service Commendation Medal he presented to BK on 4 June 1965 (Author's collection)

Notes

1. "1962 French Ambassador Letter," personal papers of Blanchard K. Watts (Author's collection).
2. Joint Service and Meritorious Service commendations, DoD file(Author's collection)
3. "1962 Pershing Missile Letter" (Author's collection).
4. "1962 Iran Earthquake Relief Letter," U.S. Dept. of State, Disaster Relief file ((Author's collection)
5. "1964 Belgrade Ambassador Letter (Yugoslavia Relief)," U.S. Embassy Belgrade (Author's collection)
6. Barbara W. Tuchman, *The March of Folly: From Troy to Vietnam* (New York: Knopf, 1984).

7. Lawrence S. Kaplan, *NATO 1948: The Birth of the Trans-atlantic Alliance* (Lanham: Rowman & Littlefield, 2007).

8. Joint Service Commendation Medal citation, awarded to Blanchard K. Watts by General Order No. 32, dated 19 May 1965, and presented on 4 June 1965 at Camp des Loges, Saint-Germain-en-Laye, France; signed by Lt. Gen. Robert L. Vittrup, USAF, Chief of Staff, U.S. Air Forces in Europe. Department of Defense award documentation. (Author's collection).

CHAPTER 33

DENOUEMENT

By late 1965, Colonel Blanchard K. Watts stepped ashore in the United States from NATO duty to take a role closer to his nature: teacher. Orders carried him to Rensselaer Polytechnic Institute in Troy, New York, where he assumed command of the Air Force ROTC detachment.[1]

The family exchanged boulevards and embassies for upstate winters. His eldest daughter enrolled across town at Russell Sage College. The younger children settled into public schools, small-town parades, and snow shoveled from the steps of brick row houses. Jeanne adapted yet again, Paris one year, Troy the next, with a grace earned across continents.

The shift appeared gentle compared to SAC's polar drills or NATO crisis councils, but it carried its own gravity. The classroom was where he could finally pour three decades of lessons into younger hands with unblinking honesty: discipline under pressure, leadership without theatrics, fidelity to principle when consequences were real.

Commandant of Cadets

At RPI, Watts became Professor of Aerospace Studies, commandant of the detachment, and mentor to cadets whose futures pointed toward Vietnam.[2] He wore the colonel's eagles, but his authority came from presence rather than insignia.

To the cadets, he was not a résumé. He was proof.

This was the officer who had flown P-40s off HMS *Archer*, strapped into jet prototypes at Muroc, carried B-52s across the polar routes, and briefed NATO generals in the grammar of deterrence. He demanded precision: drills sharp, uniforms exact, standards absolute. He never asked more than he had endured himself.

He taught by posture. Calm. Unyielding. Consistent.

For cadets barely out of adolescence, he was the steady figure in the room, the one who turned anxiety into structure. In hallways and offices, one by one, uncertainty gave way to confidence.

Family and Legacy

These teaching years were also family years. In August 1968, his oldest daughter gave birth to BK's first grandchild. The nation convulsed with war, assassinations, and unrest, but time in the Watts household measured itself in births rather than headlines.

Jeanne cradled the newborn in arms that had carried infants through Texas, North Carolina, Paris, and Troy. Recognition finally reached her as well. The Air Force issued Jeanne a Certificate of Appreciation, a document modest in form but profound in meaning.[3] It acknowledged decades of invisible labor: moves endured, children raised, careers set aside, strength carried quietly. The institution finally named what the family had always known. Her service had run parallel. Her resolve had matched his.

A Star

In those final years, quiet inquiries surfaced. Promotion. The possibility of a general's star.

Watts declined.

"I don't want to be a secretary to some general," he told his daughter and colleagues when pressed.[4] The dryness concealed conviction. Paper and politics held no appeal. Ceremony without purpose meant nothing. His measure had not changed. Teach. Mentor. Stand for airmen and cadets.

The refusal itself became part of his legacy. He remained at RPI, not from lack of ambition, but from fidelity to calling. The cadets he shaped, many bound for Vietnam's crucible, were the stars he chose to raise.

Spokane

After the final RPI class graduated in the spring of 1969, Watts accepted his last assignment. He transferred to Fairchild Air Force Base near Spokane, Washington, commanding the 92nd Civil Engineering Squadron.[6] He still called it Geiger Field, the name from his earlier era, even though the designation had changed decades before.

This was no quiet post. The squadron sustained SAC's western runway network, shelters, and alert posture. Commendations cited not flight hours but logistics, infrastructure, and readiness. Here, he became a builder, maintaining the ground upon which airpower rested. B-52s sat on the ramp restrained and waiting.

Spokane gave more than medals. Near Medical Lake, the family bought acreage on the shores of Silver Lake. Woods, water, and open ground replaced embassies and briefing rooms. Summers stretched long as grandchildren arrived by car from across the country to spend August swimming, fishing, and learning again from him what patience looked like in practice.

This lake became the family's anchor. After decades of motion, they had found stillness.

A Quiet Meritorious Service

In 1971, the Air Force awarded Colonel Watts the Meritorious Service Medal for his command at Fairchild.[5] The language was restrained: leadership, devotion, managerial excellence. Inside those words lay a summation:

"...For outstanding non-combat meritorious service as Commander, 92nd Civil Engineering Squadron, Fairchild AFB from 4 August 1969 to 31 May 1971... The singularly distinctive accomplishments of Colonel Watts culminate a distinguished career in the service of his country, and reflect great credit upon himself and the United States Air Force."

At the same ceremony, Jeanne received her Certificate of Appreciation: wife, mother, and partner acknowledged at last by the great institution both had served. Together, in the early summer of 1971, they closed three decades in uniform.

Becoming the Giant

By his retirement from military service at age fifty-two, Colonel Blanchard K. Watts had fought in 3 theaters, declined stars, closed logbooks, and handed cadets to war. Seen as a whole, he had become precisely what he once admired.

In North Africa and Italy, the giants were Cochran, Patton, and Doolittle.

At Wright Field and Muroc, Yeager, Edwards, Knerr.

In China, he taught an air force to fly jets and earned the respect and gratitude of Chiang Kai-shek.

In Paris and Bonn, the giants were Gavin, Wheeler, Smart, Elbrick.

At RPI, young men in ill-fitting uniforms looked up and saw him.

He retired a colonel, not a general. But greatness was never measured by the star on his own sleeve. It was measured by the stars others would wear. Pilots. Commanders. Cadets. Grandchildren.

History records him in orders and citations. Family remembers him in laughter carried across water. Both are true. Both endure.

On Silver Lake west of Spokane, his life found its rest and its new purpose. He was and always will be: A man who flew. A man who taught. A man who chose conviction over ambition and, in doing so, became a giant himself.

And so his story closes in Medical Lake where his life truly began again. He started out a farm boy from North Carolina who entered the Texas skies in 1941 and left behind a remarkable legacy that deserves to be remembered forever.

Notes

1. Rensselaer Polytechnic Institute, Detachment Records and

Yearbooks, 1965–1971, RPI Archives.

2. John Schlight, *The War in South Vietnam: The Years of the Offensive (1965–1968)*, USAF Air Historical Series (Washington, D.C.: Office of Air Force History, 1988).

3. "Air Force Certificate of Appreciation, Jeanne Watts" (1971), family collection.

4. Oral history interview, Judy Watts, 2008, family archive.

5. "Joint Service and Meritorious Service Awards," Dept. of the Air Force, commendation file (author;s collection

6. Dept. of the Air Force, Orders to 92nd Civil Engineering Squadron, Geiger Field, Spokane, 1971 (author's collection).

BIBLIOGRAPHY

Archival & Unpublished Sources

Blanchard K. Watts. AAF Form 5 Flight Logs and Individual Flight Records, 1942–45 and 1954–59. Blanchard K. Watts Papers, Author's collection.

Blanchard K. Watts. Handwritten War Diary, Jan.–Mar. 1944; Handwritten Diary and AAF Form 5 Flight Logs (May–June 1944). Blanchard K. Watts Papers, Author's collection.

Blanchard K. Watts. Pilot Flight Logbook (auxiliary reference, Jan.–Feb. 1944 hops). Blanchard K. Watts Papers, Author's collection.

Watts, Blanchard K. The Rise and Demise of the Joker Squadron. Unpublished memoir, ca. 1983. Reproduced in Appendix A. Author's collection.

Headquarters Chinese Air Force. General S. M. Wang to Lt. Col. Blanchard K. Watts, through Colonel Edward F. Rector, Chief, Air Section, MAAG, Taipei, Taiwan. "The Presentation of a CAF Pilot's Wings," 7 September 1953. Translation. Blanchard K. Watts Papers, Author's collection.

Judy Watts. Oral history interview, 2008. Author's collection.

Rensselaer Polytechnic Institute (RPI). Air Force ROTC Detachment Records and Yearbooks, 1965–71. RPI Archives, Troy, NY.

United States Department of the Air Force. Personnel Orders, April 1958, appointing Blanchard K. Watts to the Regular Air Force. Blanchard K. Watts Papers, Author's collection.

United States Department of the Air Force. Orders, January 1959, promoting Blanchard K. Watts to Colonel and assigning him to Maxwell AFB / 3800th AB Wing. Blanchard K. Watts Papers, Author's collection

United States Department of the Air Force. "Air Force Commendation Medal Citation, approved March 1960." 1959_air_force_commendation_medal.pdf. Blanchard K. Watts

Papers, Author's collection.

United States Department of the Air Force. Joint service and Meritorious Service commendation files. "Joint Service and Meritorious Service Awards." joint_service_and_meritorius_servce.pdf. Blanchard K. Watts Papers, Author's collection.

United States Department of the Air Force. Orders to 92nd Civil Engineering Squadron, Fairchild AFB, Spokane, WA, 1971. Blanchard K. Watts Papers, Author's collection.

United States Department of State. "1962 Iran Earthquake Relief Letter." 1962_iran_earthquake_letter.pdf. Blanchard K. Watts Papers, Author's collection.

United States Embassy Belgrade. "1964 Belgrade Ambassador Letter (Yugoslavia Relief)." 1964_begrade_ambassador_letter.pdf. Blanchard K. Watts Papers, Author's collection.

"1962 French Ambassador Letter." 1962_french_ambassador_letter.pdf. Blanchard K. Watts Papers, Author's collection.

"1962 Pershing Missile Letter." 1962_pershing_seargant_germant_letter.pdf. Blanchard K. Watts Papers, Author's collection.

"Air Force Certificate of Appreciation, Jeanne Watts." 1971. Author's collection.

"Mrs. Jeanne Watts Is Company's First Graduate Lady Engineer." The Spotlight (Carolina Power & Light Company employee newsletter) 14, no. 11 (May 1951): 7.

Books and Scholarly Works

Ambrose, Stephen E. D-Day, June 6, 1944: The Climactic Battle of World War II. New York: Simon & Schuster, 1994.

Blumenson, Martin. Patton: The Man Behind the Legend, 1885–1945. New York: William Morrow, 1985.

Chennault, Claire L. Way of a Fighter. New York: G. P. Putnam's Sons, 1949.

Craven, Wesley Frank, and James Lea Cate. The Army Air Forces in World War II. Vol. 5, The Pacific: Matterhorn to Nagasaki. Chicago: University of Chicago Press, 1953.

Cullum, George W. History of the Army Air Forces Service Command. Washington, DC: USAF Historical Division, 1945.

Fall, Bernard. Hell in a Very Small Place: The Siege of Dien Bien Phu. New York: Da Capo Press, 1966.

Ford, Daniel, with Glen Edwards. Glen Edwards: The Diary of a Bomber Pilot. Washington, DC: Smithsonian Institution Press, 1998.

Gaddis, John Lewis. Strategies of Containment: A Critical Appraisal of American National Security Policy during the Cold War. New York: Oxford University Press, 1982.

Howe, George F. Northwest Africa: Seizing the Initiative in the West. Washington, DC: U.S. Army Center of Military History, 1957.

Hull, William. The Army Air Forces Depot System in World War II. Maxwell Air Force Base, AL: Air Force Historical Research Agency, 1952.

Kaplan, Lawrence S. NATO 1948: The Birth of the Transatlantic Alliance. Lanham, MD: Rowman & Littlefield, 2007.

Kohn, Richard H. Eagle and Sword: The Federalists and the Creation of the Military Establishment in America, 1783–1802. New York: Free Press, 1975.

LeMay, Curtis E., with MacKinlay Kantor. Mission with LeMay: My Story. Garden City, NY: Doubleday, 1965.

Maurer, Maurer. Aviation Cadet Training in World War II. Maxwell Air Force Base, AL: USAF Historical Studies, 1967.

Maurer, Maurer, ed. Combat Squadrons of the Air Force, World War II. Washington, DC: U.S. Air Force Historical

Division, 1969.

McGowan, Sam H. The Hump: America's Strategy for Keeping China in World War II. Annapolis, MD: Naval Institute Press, 2016.

Morison, Samuel Eliot. History of United States Naval Operations in World War II. Vol. 2, Operations in North African Waters, October 1942–June 1943. Boston: Little, Brown, 1947.

———. History of United States Naval Operations in World War II. Vol. 9, Sicily–Salerno–Anzio. Boston: Little, Brown, 1954.

———. History of United States Naval Operations in World War II. Vol. 11, The Invasion of France and Germany. Boston: Little, Brown, 1957.

Nelson, Craig. Pearl Harbor: From Infamy to Greatness. New York: Scribner, 2016.

Poole, Walter S. Adapting to Flexible Response, 1960–1968. Washington, DC: Office of the Secretary of Defense, 2013.

Prange, Gordon W. At Dawn We Slept: The Untold Story of Pearl Harbor. New York: Penguin, 1981.

Romanus, Charles F., and Riley Sunderland. Stilwell's Mission to China. United States Army in World War II: China–Burma–India Theater. Washington, DC: Government Printing Office, 1953.

Schlight, John. The War in South Vietnam: The Years of the Offensive, 1965–1968. U.S. Air Force Historical Series. Washington, DC: Office of Air Force History, 1988.

Suid, Lawrence H. Guts and Glory: Great American War Movies. Washington, DC: Addison-Wesley, 1978.

Tuchman, Barbara W. Stilwell and the American Experience in China, 1911–45. New York: Macmillan, 1970.

———. The March of Folly: From Troy to Vietnam. New York: Alfred A. Knopf, 1984.

Twining, Nathan F. Neither Liberty nor Safety: A Hard Look at U.S. Military Policy and Strategy. New York: Holt, Rinehart and Winston, 1966.

U.S. Army Air Forces. Pilot Training in the AAF, 1939–1945. Army Air Forces Historical Office, 1946.

U.S. Army Air Forces. The Air Phase of the North African Invasion, November 1942–February 1943. AAF Historical Study No. 5. Washington, DC: Army Air Forces, 1944.

U.S. Army Center of Military History. Advisory Efforts in Vietnam, 1955–60. Washington, DC: U.S. Army Center of Military History.

U.S. Army Center of Military History. Pearl Harbor: 7 December 1941. CMH Pub 72-7. Washington, DC: U.S. Army Center of Military History.

U.S. Army Center of Military History. Sicily and the Surrender of Italy. CMH Pub 72-16. Washington, DC: U.S. Army Center of Military History.

Wilson, John B. Maneuver and Firepower: The Evolution of Divisions and Separate Brigades. CMH Pub 60-14. Washington, DC: U.S. Army Center of Military History, 1998.

Government & Official Reports (Non-Book)

"Castle Bravo Report, Bikini Atoll, 1 March 1954." U.S. Atomic Energy Commission.

Cooper, Robert T. "The 33rd Fighter Group: 'Fire from the Clouds'." Research Report 88-0600. Air Command and Staff College, Air University, Maxwell Air Force Base, Alabama, 16 February 1988.

NASA History Division. "Sputnik and the Dawn of the Space Age." October 1957.

National Security Council. "Report to the President by the Security Resources Panel of the Science Advisory Committee (Gaither Report), 7 November 1957." In Foreign Relations of the United States, 1955–1957, vol. 19, National Security Policy; Information Policy; Arms Control and Disarmament. Washington, DC: U.S. Government Printing Office.

United States Department of State, Office of the Historian. "The Suez Crisis, 1956."

United States Department of State, Office of the Historian. "The Hungarian Revolution of 1956."

United States Department of State, Office of the Historian. "The Eisenhower Doctrine and the 1958 Lebanon Crisis."

United States Navy, Naval History and Heritage Command. "Pearl Harbor Attack, 7 December 1941: Summary of U.S. Ships Damaged and Aircraft Lost." Accessed August 2025.

Reference to Specific Unit/Operational Documents

33rd Fighter Group Combat Digest. Reunion booklet, 1984.

Allied Air Forces Liaison Report. "P-47 Tactical Exchange," 25 January 1944. Air Force Historical Research Agency microfilm.

Army Air Force Historical Research Agency (AFHRA). Unit combat digests, In Memoriam rolls, and official orders related to 33rd Fighter Group and associated units. Maxwell Air Force Base, AL.

Civil Rights, Politics, and Context

Federal Judicial Center. "Civil Rights and Federal Courts: Montgomery and the Bus Boycott, 1955–56."

Library of Congress. Civil Rights History Project.

Roosevelt, Franklin D. "Day of Infamy Speech." 8 December 1941. Congressional Record.

"Final Declaration of the Geneva Conference." 21 July 1954

APPENDIX A

THE RISE AND DEMISE OF THE JOKER SQUADRON

The following pages reproduce Col. Blanchard K. "BK" Watts's original memoir of the Joker Squadron events, written circa 1983. Pagination, wording, and punctuation appear as in the surviving manuscript.

THE RISE AND DEMISE

of the

JOKER SQUADRON

BY

BLANCHARD KING WATTS

COLONEL (RET) USAF

Most everyone knows of Doolittle's "30 Seconds Over Tokyo" raid in April of
1942, but few know of Cochran's JOKER squadron of thirty-six P-40F aircraft
catapulted from the British escort carrier HMS Archer off the coast at
Casablanca in November of the same year. Probably no one wanted to publicize
an event in which a Major led thirty-five "shavetail" 2nd Lieutenants, with
no training whatsoever, on a catapult launch. In fact, none had ever seen
a catapult until on board the Archer.

I hadn't thought much about that event in recent years until asked to put my
recollections on paper. It has been almost forty-two years since I was
propelled out over the bow of that ship, like a toy plane being shot by a
rubber band. While the operation created lasting impressions, some have
become larger than life, while others have faded with time. I hope to jog
my memory with the aid of my Air Corps Form 5 Flight Record, my personal
pilot's log book with scattered notes, various photographs, travel orders
and other bits of information.

One of the most unusual aspects of the mission was the manner in which the
planes and pilots were assembled and loaded on the ship. None had the
slightest hint of what was happening until actually on board. I was one of
the "peas" in the pod of 2nd Lieutenants who were gathered from units along
the East coast, all in the same manner.

After graduating from flying school at Kelly Field, Texas, on 2 July 1942, I
was assigned to the 315th Fighter Squadron, 324th Group, flying P-40's at
Philadelphia Municipal Airport. On 26 October 1942 I had acquired 108 hours
in the P-40, when I was called to squadron headquarters along with four other
2nd Lieutenants, and given a copy of orders and a train ticket. The one-
sentence order was to "proceed to Middletown Air Depot, Pa., for the purpose
of testing and ferrying military aircraft." There was nothing Unusual about
this. Pilots were used to testing and ferrying aircraft from a depot to
active flying units on a routine basis. I was on a train within an hour,
and at Middletown within the next hour. The following morning I was assigned
a P-40F. My instructions were to test-hop it, and if the aircraft was O.K.,
not to land back at Middletown, but to continue to Floyd Bennett Field, N.Y.

After landing there, I was told to leave the aircraft, go to Mitbhell Field,
N.Y., by vehicle and await further instructions. At Mitchell I found a
number of 2nd Lieutenants who had gone through the same ferrying mission,
In facg, by the next day the number had swelled to thirty-five. We now began
to smell something ou the wind, but any concern over future unpleasant events
was quelled by daily passes into New York City, each with a deadline by which
to check back in at Mitchell. With freedom to roam New York, and a finance
office available, the only worry about tomorrow was whether we would get
another pass. On 1 November 1942 there were no passes, but instead we were
loaded in vehicles and proceeded to the New York harbor. Approaching the
harbor, we saw a flat-top, her deck covered from stern to bow with our P-40's.
We must have been the last to board, because the ship was underway almost
immediately. At sea, cut off from mailboxes and telephones, we could now be
informed of our upcoming mission. We met and were briefed by our new leader
and Commanding Officer, Major Phillip G. Cochran.

Those of my generation who read cartoonist Milton Caniff's comic strips of
"Terry and the Pirates" and "Steve Canyon" can imagine the existance and
comedy 6f the JOKER squadron. Caniff portrayed real people in his strips,
including appearance, phrases and mannerisms in his characters. (I met the
Dragon Lady, an Air Corps widow in real life, in Dayton, Ohio, and knew who
she was before being told.) Cochran was Flip Corkin in "Terry and the
Pirates," and General Phillerie in "Steve Canyon." The entire mission of the
JOKER squadron could have been an episode in one of the comic strips, but our
mission was not imagination or fiction, although comical at times. It was
real --- I was there!

Back to Cochran's briefing --- the Allied invasion of North Africa was underway. D-Day was 8 November 1942. The main invasion force was already at sea,
and grouping to hit North Africa at Algiers, Oran and Casablanca. Our convoy
was not to be part of the initial landing force, but rather a small convoy of
fast ships capable of avoiding and/or out-running the U-boats. Our planes
and pilots had been assigned to the 33rd Fighter Group as replacements for
anticipated losses. (The 33rd was to launch off Casablanca on D-Day from
the USS Chenango.)

The first part of the briefing seemed routine enough, with the pilots
somewhat inattentive at times. We had flown the planes into Floyd Bennett,
and while we were living it up in downtown New York, they had been towed to
the harbor and loaded onto the flat-top by crane. We came aboard, and the
ship weighed anchor. So --- we would cruise into Casablanca harbor and go
through the reverse in unloading and towing to some airstrip. Not so!
Cochran gained our full attention when he announced we would be catapulted
from the ship while at sea, a real attention line!

We were on board a small British carrier, HMS Archer. Her normal mission
was to act as convoy escort carrier, and was equipped with Swordfish torpedo
planes. These were light bi-planes, capable of launch by catapult, and
recovery on the short deck. On this trip, the P-40's covered the entire
deck, the Swordfish were in the hangar below deck, with no way to bring them
up for launch, let alone recovery. Even if the P-40's were launched,
recovery for them by ship was not possible.

We were not yet an organized unit, but by the end of the briefing Cochran
had his organization. Since we were all 2nd Lieutenants, he simply pointed
a finger and asked your name. In this way he appointed the various officers
of administration --- operations, maintenance, medical and so on. When his
finger pointed at me, Maintenance Officer came up. Someone asked if this was
a joke, and there we had the JOKER squadron.

During the next thirteen days I learned more about the P-40 than any school
could have taught. In fact, all the pilots did, and they learned from the
British mechanics and a half-dozen U.S. enlisted mechanics, on board as
support personnel. The demand for those few mechanics was so great that they
became advisors, and the pilots did the work. Each pilot was his own
mechanic and crew chief, with a real incentive to have his aircraft in
shape at launch time. We were told that when our plane came up for launch, it
would simply be pushed overboard if the launch officer felt it was not in
shape for launch. Maintenance was not easy. The aircraft were lashed down
on deck so close that propellers could not be turned over for lack of clearance
from other aircraft wings, control surfaces or fuselage. No way to run up

engines except by releasing tiedowns and tediously shifting aircraft enough
for propeller clearance. This could only be done during calm seas, and with
a lot of manpower. We had the manpower, but not always a calm sea. With the
aircraft sitting on deck in salt spray for two weeks, there was no option but
to run engines sometime prior to launch, or face the possibility of pushing
most of the aircraft over the side. I can still remember the calm seas along
about the eleventh day, when we spent all day shifting aircraft and running
engines. Cleaning the six 50-caliber guns was no picnic. Prior to the
installation of the guns at the depot, they were dipped in cosmoline, a thick,
greasy lubricant, for protection. Normally, at an established base, the
ordnance branch has vats with solvents for such cleaning. There were no vats
on HMS Archer, and, worse still, the weather most of the time was such that we
could not clean the guns topside. This required removal, taking the guns
below to the hangar deck, where we couldn't even use gasoline or other common
solvents. Days and days of wiping the guns, inside and out, with rags, paper
or whatever we could scrounge from the ship. Needless to say, the British
were cleaned out of such materials.

Aircraft weight was extremely critical because the P-40 exceeded the design
capability of the catapult. This meant reducing weight by every pound
possible. The aircraft were stripped of routine items like tool kits, nose
plugs, tie-downs, etc. (you wouldn't believe!) We went so far as to unload
four of the six guns, and reduce the remaining two guns to fifty rounds each.
After computing flying time from the anticipated launch position to shore, we
reduced our gasoline to bare minimum to get direct to the landing strip, with
no reserve for fighting or straying off course en route. With all this, we
still exceeded the catapult design capability, which was of much concern to the
launch officer.

The only source for training was the British --- the pilots and the launch
officer. The Swordfish pilots had never flown the P-40, and consequently
could only guess how it would react during and immediately following the
catapult release. It was the same with the launch officer. During the daily
briefings, discussions and dry runs, there was always the difference in aircraft characteristics. We tried to anticipate exactly how the P-40 was

going to react. The British were much more concerned about this than we
wefe. With our cocky confidence, we had this attitude --- you launch us,
and we'll fly!

Normally, thirteen days would be considered a long trip from New York to
Casablanca, but among the pilots there were mixed feelings. On the one
hand, launch day came much too soon, because we knew we weren't prepared,
and a bit of dread crept in under the confident attitude when we thought of
being rolled up for the actual launch. On the other hand, we had been
cruising off the African coast for several days, waiting for information of
a secure landing field, and the German U-boats were beginning to collect like
vultures. General quarters alarm was a common occurence throughout the trip,
but now it had become far too frequent. A common sight now became the
geysers from depth charges from our escorting destroyers. There was very
little information coming from ashore, and a nervous atmosphere began to
develop aboard ship. We did learn the 33rd Gp. P-40's from the USS Chenango
had landed on a field at Port Lyautey, some distance north of Casablanca.
Obviously there was more resistance in the Casablanca area than anticipated.
We finally got word on 13 November that the field at Casablanca was secure
for our landing. On the morning of 14 November we began our launch.

Major Cochran had divided the thirty-six aircraft into two flights of
sixteen, plus four as spares. He was to lead the first flight, and appointed
me as leader of the second. At the start of the launch, all pilots were in
aircraft, with radios on. Engines were started as soon as the aircraft ahead
moved enough for propeller clearance. Pilots close enough were to observe the
launches ahead to gain all knowledge possible. Also, after Cochran's launch,
he was to pass his reactions back to all pilots by radio. Cochran was
launched, but the catapult broke down because of overload. With minimum fuel
in the tanks, he could circle only momentarily before being forced to head
for shore. Among other instructions, he moved me from the leader of the
second flight to his position in the first flight, if and when the catapult was
again operational. This meant my changing aircraft from the one I had nursed,
cleaned and become closely attached to for thirteen days. There wasn't much
time to worry, for about the time I got buckled in the seat, the signal came

to start engine. I still remember a feeling of not being ready for launch.
I had anticipated slowly moving up the line, as aircraft were launched, of
doing a final engine run-up, observe launches ahead, get the heading to shore
from the board behind the launch officer, and a generally slower-moving procedure. I was suddenly in a different aircraft, with no time to check out the
cockpit, run engine or get positioned for the boot in the rear that was about
to come. The launch officer was urgently twirling his flag for me to "rev"
up to full throttle. Looking past him, I saw the big 80 degrees (the heading
to shore) just as his flag came down. As the plane headed out over the ship's
bow, a bit of Cochran's radio instruction flashed through my mind --- "don't
stall!" He had emphasized that we would be at or below flying speed as we were
released from the catapult. This flash no doubt kept me dry. The plane was
shuddering and buffeting in a stall, and sinking toward the water below. Without Cochran's warning, the reaction would have been to hold back-pressure on
the stick to stop the sink. I let pressure off, and just when it seemed the
ocean swells were going to swallow me, the buffet stopped, I could feel the
bite of the propeller push me back in the seat, and I truly felt the thrill of
going into the "wild blue yonder." I cleaned up and trimmed the bird, and
made the wide circle to allow follow-on planes to join up. None joined. I
came back across the ship, and knew how lonely Cochran must have felt a short
time earlier. The catapult was down again! There was no estimate as to the
next launch. I made another pass across the HMS Archer in a final farewell,
before heading for Casablanca.

I have since been in a lot of unusual situations, but have never again had the
kind of feeling as when the convoy of ships faded behind, and there was nothing
but blue water and a few scattered clouds ahead. You know land is out there
somewhere, but because of scant information on board ship, you have no idea of
what to expect when reaching that land. You are not even sure the designated
airstrip is in friendly hands. And you are all alone. The feeling would have
been a lot different with even one other plane out there in formation. It
seemed like a very long 30-minute flight. I checked and re-checked everything
a dozen times. Of special concern was compass heading. Screw up on direction,
and you simply disappear at sea. But 80-80 was burned in my brain. Before
launch we were told we were approximately 80 miles from shore, and when I saw

the 80 degrees on the board the instant before launch, there was no for-
getting. Yet, as I strained my eyes for a shoreline, I kept seeing
shorelines come and disappear like mirages, and began to worry about the
compass. The only cross-check was the sun, which I kept "eye-balling"
and knew I was heading in an easterly direction, which would definitely
take me to land, even if not where I wanted to be. Finally, a mirage
showed up that stayed put, but to replace that concern was that of enemy
territory, with the possibility of fighters, anti-aircraft fire and the
problem of finding the landing strip before running out of fuel. The
compass was correct, I hit shoreline, turned right as planned, and there
lay Casablanca. Avoiding the city and harbor, I found the airfield. There
was not only one P-40, but many, parked on the edge of the strip. I had
expected to see Cochran's plane, but was surprised to find others. Now the
loneliness disappeared, the cockiness returned, and I did the typical
fighter-pilot low-approach landing by buzzing the strip, pulling up,
chopping the throttle, doing a 360 degree overhead and plunking it down on
the grass.

Cochran was frantic for information about what was happening on the HMS
Archer. Of course, all I could tell him was a repeat of what had happened
to him. We paced the strip all afternoon, sweating planes in. At first,
there were flights of one and two, then as many as seven or eight, but
never an orderly-appearing operation. Finally, there were thirty-three
of the thirty-six on the airstrip. We learned from pilots late in launch
that two had gone into the sea after release, and the pilots had been
picked up by destroyers. The other had launched successfully, but pilot
and plane had simply disappeared.

The 33rd Group was moving from Port Lyautey to Casablanca, and the strip
was becoming crowded. This gave Cochran an excuse to hang on to his pilots
and planes a while longer. He learned of a French P-36 (Mohawk) squadron
at Rabat which was no longer operational after the invasion. Since the P-36
was much the same as the P-40, except for the engine, he knew there would
be at least some parts and logistical support there. Two days after landing
at Casablanca, we were at Rabat. Even though the French were technically

our enemies a few days earlier, Cochran's tact, diplomacy and hypnotic approach brought complete support from the unit. Here Cochran formally organized his JOKER squadron, set up a firing range on the beach and started training his pilots for combat. Everyone knew we were assigned to the 33rd Group, and this lark could not last. In the meantime, we were living it up with the French, flying, and far removed from the rigid rules and discipline of the Army.

Colonel Momeyer, of the 33rd Group, flew up about three weeks later, and must have told Cochran to cut the dreaming, and bring the planes and troops back to the 33rd. Cochran still would not give up. The next day, on 7 December, he took me and one other pilot, and the three of us headed for 12th AF Headquarters at Oran. This was evidently an attempt to salvage something, or work out some other scheme. Whatever was worked did not include us two Lieutenants. The following day we were put on a transport and sent to Casablanca. Our ego was completely blown because we were not allowed to fly our P-40's back! It was like taking the horse from a cavalry soldier.

At Casablanca we found the JOKER pilots and planes, but no JOKER squadron. We had been split three ways, and assigned to the squadrons of the 33rd Group. If the JOKER squadron even existed at all, it was from 1 November to 7 December, 1942. Although never recognized for it, these pilots did accomplish an unusual feat. They faced the adversities of minimum overall flying experience, no knowledge of catapult launch (not even a briefing from anyone who had launched in a P-40 from ship or land) and preparing their individual planes after two weeks in salt spray. Yet, they accomplished what should be considered a highly successful mission --- they launched thirty-six P-40's from ship and moved them to shore, with the loss of one pilot and three aircraft. Considering the circumstances, a well-organized and experienced unit probably would not have done as well. Maybe it is because his confidence outweighs his common sense that a newly-commissioned 2nd Lieutenant, eppecially a newly-commissioned fighter pilot, will tackle the impossible, given the responsibility. Add an abundance of luck, and he will be around to brag about it later.

Note: When the 33rd moved to the Tunisian front at Thelepte air strip the
first part of January, 1943, Cochran was sitting there with his one P-40,
claiming he was the Base Commander, and would run the show at his base (or
something like that.) He flew with the 33rd for a short while, getting
involved in some unusual combat missions, before returning to the States.
I passed through Burma in 1944 and learned he was in command of an Air
Commando unit made up of fighters, light bombers, cargo, gliders and
probably a few water buffalo as logistical support. It was the exact
situation in which one would expect to find Caniff's "Flip Corkin."

JOKERS of the 33rd Ftr. Group who launched from HMS Archer.

Major   Cochran, Phillip G.   0-22464

2nd Lts. Adams, Clyde P.   0-790605
          Arterburn, Haskell E.   0-661177
          Bishop, Lynn E.   0-661180
          Brock, Lester W.   0-791066
          *Carpenter,
          Cibel, Harvey J.   0-791071
          Conner, Robert J.   0-659666
          Cottrell, Charles L.   0-790633
          Cross, Willard D.   0-724390
          Cuthbert, Richard R.   0-412620
          Duncan, Berry J.   0-790640
          French, Harry E.   0-727438
          Gardner, Shirley E., Jr.   0-659690
          Goulait, Bert J.   0-349111
          Haines, Harry R.   0-661310
          Hawke, Jack E.   0-728655
        **Hearing, Oscar V.   0-728657
        **Kantner, Robert P.   0-728676
          Keller, Archie C.   0-727479
          King, Charles E.   0-791883
          Matuch, George, Jr.   0-433565
          McBride, William P.   0-727506
          McCreight, Claude E.   0-728694
          McMills, Phil R.   0-727516
          Murdock, Richard D.   0-727524
          Nightingale, George W.   0-791027
          Rodriguez, Leopoldo V.   0-664661
          Sayles, Herbert C.   0-791149
          Silberstein, Irving S.   0-791152
       ***Smith, Robert
          Thomas, Tom A., Jr.   0-664691
          Thompson, Lassiter   0-661251
          Tobin, Edward J.   0-791167
          Walker, John S.   0-661409
          Watts, Blanchard K.   0-661255

  * Lt. Carpenter launched successfully, but did not show up at Casablanca.

 ** Lts. Kantner and Hearing stalled on launch, and went into drink. Both
     were picked up by destroyers.

*** Lt. Smith was killed at Rabat before a complete roster was compiled.

9 November 1942 - - - P-40's "stacked" on deck of
HMS Archer. After fourteen days on deck in rain,
wind and salt spray, they still launched.

Lt. Smith launches from HMS ARCHER on 14 Nov 42 off coast at Casablanca.

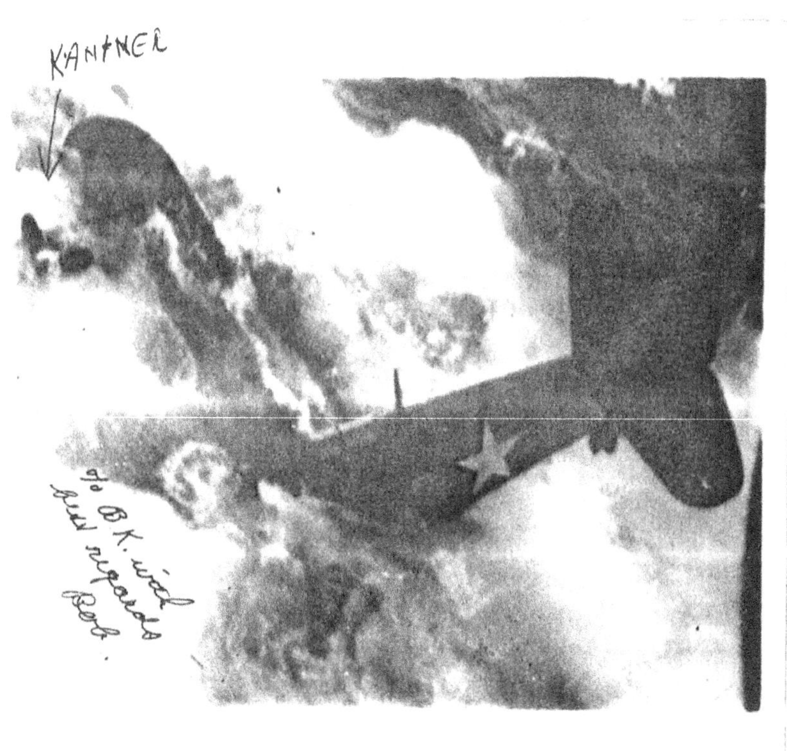

Lt. Kantner was not successful, and took a swim. Note that he was able

to autograph the photo.

Above: Momeyer (left) and Cochran
at Rabat O/A 1 Dec 42.
After this visit by Momeyer,
Cochran left for Oran.

Left: Lt. Watts (author?) doing
maintenance duties at Rabat.

16 December, 1942.

SPECIAL ORDER)
                :
NO.   221)

E X T R A C T

1. The following named Officers of the "J" Squadron, 33rd Fighter Group, are relieved from assignment and duty thereof, and assigned to the organizations indicated, effective this date.

58th Fighter Squadron

2nd Lt.	LEOPOLDO N. RODRIGUEZ,	0-664661,
2nd Lt.	TOM A. THOMAS, JR ,	0-664691,
2nd Lt.	LASSITER THOMPSON,	0-661251,
2nd Lt.	JACK E. HAWIE,	0-728655,
2nd Lt.	LYMAN R. BISHOP,	0-661180,
2nd Lt.	LESTER W. BROCK,	0-791066,
2nd Lt.	BERT C. POLLAIT,	0-349111,
2nd Lt.	IRVING S SILBERSTEIN,	0-781152,
2nd Lt.	ROBERT P. KANTNER,	0-728676,
2nd Lt.	RICHARD R. CUTHBERT,	0-412620.

59th Fighter Squadron

2nd Lt.	EDWARD J. TOBIN,	0-791167,
2nd Lt.	RAYMOND A. WATTS,	0-661255,
2nd Lt.	OSCAR V. HEARING,	0-728657,
2nd Lt.	HASKELL B. ARTERBURN,	0-661177,
2nd Lt.	BERRY J. DUNCAN,	0-790640,
2nd Lt.	HARRY E. FRENCH,	0-727438,
2nd Lt.	HARRY R. HAINES,	0-661310,
2nd Lt.	CLAUD E. McCREIGHT,	0-728694,
2nd Lt.	GEORGE W. NIGHTINGALE,	0-791027,
2nd Lt.	CLYDE P. ADAMS,	0-790605,
2nd Lt.	CHARLES L. COTTRELL,	0-790633.

60th Fighter Squadron

1st Lt.	GEORGE M. TUCH, JR.,	0-433563,
2nd Lt.	HARVEY J. CIBEL,	0-791071,
2nd Lt.	ROBERT J. CONNER,	0-659666,
2nd Lt.	WILLARD D. CROSS,	0-724390,
2nd Lt.	SHIRLEY E. GARDNER, JR.,	0-659690,
2nd Lt.	WILLIAM F. McBRIDE,	0-727506,
2nd Lt.	PHIL L. McMILLS,	0-727516,
2nd Lt.	ARCHIE C. KELLER,	0-727479
2nd Lt.	JOHN S. WALKER,	0-661409,
2nd Lt.	CHARLES R. KING,	0-791883,
2nd Lt.	RICHARD D. MURDOCK,	0-727524.

*   *   *

By order of Lt. Colonel MOMYER:

OFFICIAL:

William J. Hall
WILLIAM J. HALL,
Captain, Air Corps,
Adjutant.

WILLIAM J. HALL,
Captain, Air Corps,
Adjutant.

DISTRIBUTION:
1 - Orig to file;
1 - CG, WTF;
1 - CO, Cazes AB;
1 - CO, 7th Fighter Wing;

APPENDIX B

Awards and Citations

This list reflects the decorations, medals, badges, and unique commendations accumulated by Colonel Blanchard K. Watts over his three decades of distinguished service. From North Africa in World War II through the early Cold War and Korea, and his later diplomatic and advisory roles.

Decorations, Medals, and Badges

USAF Decorations

Command Pilot Wings (U.S. Air Force)
Distinguished Flying Cross
Meritorious Service Medal
Air Medal with 12 Oak Leaf Clusters
Joint Service Commendation Medal
Air Force Commendation Medal
Presidential Unit Citation - 33RD Fighter Group

U.S. Campaign and Service Medals

American Defense Service Medal
American Campaign Medal
European–African–Middle Eastern Campaign Medal
China–Burma–India Theater Campaign Medal
World War II Victory Medal
National Defense Service Medal
Korean Service Medal
Armed Forces Reserve Medal
Air Force Longevity Service Ribbon

Marksmanship and Qualification

Small Arms Sharpshooter Ribbon/Badge

Foreign Awards

Chinese Nationalist Air Force Wings, serial no. 72047

Selected Letters of Commendation and Special Recognitions

Commendation for exceptional ability and tact in Paris, 1962
(from General Bondley, Lt. Gen. James M. Gavin, and others)
Commendation for flawless cooperation with American and

German missile units, Germany, 1962 (from General Bondley, Lt. Gen. G. B. Russell, General Earle G. Wheeler, Generalmajor Kuntzler)

Commendation for diligence and efficiency in Iran earthquake relief, 1962 (from General Bondley, Admiral John C. Hayden, Robert L. Campbell)

Commendation for instrumental role in humanitarian relief during Skopje earthquake and Zagreb floods, Yugoslavia, 1964 (from Frank Rouse, Gen. Jacob E. Smart, Ambassador C. Burke Elbrick)

Japanese Officer's Sword, conveyed by the Chinese Air Force (CAF), 1953

WATTS
U.S.
U.S.

INDEX

1st Infantry Division 67
33rd Fighter Group 27, 28, 38, 91, 117, 120, 143, 153
34th Infantry Division 69, 70
42nd Air Division 207
58th Fighter Squadron 37
59th Fighter Squadron 37, 104, 110, 112
60th Fighter Squadron 37, 45, 118
334th Division (German) 69
522nd Squadron 203

A

A-20 83, 85
Agadir 50–51
aircraft fatalities 27, 30, 45–46, 63, 85, 94, 105, 121, 124–125, 157
Air Medal 53
  12th Oak Leaf Cluster 114
Alexander, Gen. Sir Harold 52
Anderson, Kenneth, Gen. 74
Anzio 123
Archer (HMS) 11–12, 17–18

B

B-24 Liberator 150
B-25 83, 109
B-26 54
B-52 Stratofortress 227–229
Barnes, Pancho 185
Barney Dean 99
Bergstrom AFB 201
Berteaux 54
Bizerte 74–76
Bob Hope 99–100
Bondley, Charles J., Maj. Gen. 234–235
Bong, Richard, Maj. 183
Bradley, Omar N., Gen. 52, 62, 66, 73–74, 84
Burcham, Milo 182

C

C-47 Gooney Bird 89, 94, 132
C-54 Skymaster 135
C-78 Bobcat 126

Caniff, Milton 12
Caproni Ca.100 95
Carolina Power & Light 197–198
Carpenter, Marvin, 2LT 25
Casablanca 23, 35
Cassino 113
Cellini, Oliver, Col. 153, 158
Cercola 118
Chenango (USS) 17
Chennault, Claire Lee, Gen. 137
Chiang Kai-shek 142, 209
  Madame Chiang 142
Chilstrom, Kenneth O., Capt. 183, 190
Clark, Mark, Gen. 103
Cochran, Philip G., Maj. 11–12, 18, 27–28
Convoy UGF-2 11

D

Distinguished Flying Cross 100
Distinguished Unit Citation 44
Doolittle, James, Gen. 25–26, 52–53

E

Edwards AFB. See Muroc Army Air Field
Edwards, Capt. Glen W. 184, 190
Eisenhower, Dwight D., Gen. 16, 83, 100, 203
Elbrick, Ambassador C. Burke 236
El Guettar 58, 61
European Command (EUCOM) 233–237

F

F-84 Thunderjet 207
Fairchild AFB 241
Fedala 16
Fifth Panzer Army 43
Floyd Bennett Field 10
Flying Tigers 141
Fourteenth Air Force 145
Frances Langford 99–100
Fredendall, Lloyd, Gen. 16, 52
Fw 190 95–96, 105, 118, 121

G

Gavin, James M., Gen. 234
Gilkeson, Adlai H., Brig. Gen. 141, 152, 161
Great Falls AFB 215

H

Hal Block 99
Halliday, Robert, Maj. 135
Hayden, John C., Adm. 235
Hearing, Oscar, 2LT 24
Hicks Field 1–3
Hill 609 (Djebel Tahent) 65–66
Himalayas 136

I

II Corps 69
Iran Earthquake (1962) 235

J

Jean Bart (ship) 23
Joint Service Commendation Medal 236
JOKER Squadron 12, 17, 26, 28–30, 45–46, 147

K

Kantner, Robert, 2LT 24
Karachi 134–135
Kelly Field 5–7
Kesselring, Albert, Field Marshal 90, 111
Knerr, Hugh J., Gen. 181
Kunming 137

L

LeMay, Curtis E., Gen. 217
Lockheed 182–183

M

Maknassy 44–45
Malmstrom AFB. See Great Falls AFB
Malmstrom, Einar A., Col. 217
Mateur 67, 74
Me 109 (Messerschmitt) 54, 121
Mediouna Airfield 20, 23, 25
Mehdia 16

Meritorious Service Medal 241
Mitchel Field 11
Momyer, William, Col. 27–30, 37, 91
Montgomery, Bernard L., Field Marshal 43, 62, 71, 73–74, 93, 97, 111
Moore, Benjamin, SSgt. 5, 140
Moyer, Harry, Capt. 139–140
Mt. Vesuvius 118
Muroc Army Air Field 174–175, 178–179, 181–183, 185–186

N

Naples (Napoli) 109
Nettuno 124

O

Operation AVALANCHE 103
Operation HUSKY 87–88
Operation MATTERHORN 153
Operation TORCH 15

P

P-39 Airacobra 44
P-40 Warhawk
  P-40D 151
  P-40F 9, 13, 17–20, 23, 36–37, 44
  P-40L 53, 67, 70, 87, 91, 95, 100–101, 106, 112, 117, 125
  P-40N 138
P-47D 146, 151, 158–159
P-51 Mustang 150
P-80 Shooting Star 174, 178, 182–186
Paestum 107, 109–110
Palais de l'OTAN 233
Pantelleria 83–85
Patton, George S., Gen. 35–36, 52–53, 57, 66, 90–91, 96, 99–100, 168–169
Pearl Harbor 5
Pershing Missile 234
Pont du Fahs 66–67
Port Lyautey 16

Q

Quesada, Elwood, Col. 11

R

Rabat 25, 28, 35–36

Randolph Field 4–5
Reed, Bernard "Bernie", SSgt. 39, 106–107
Rensselaer Polytechnic Institute 239
Rommel, Erwin, Field Marshal 43, 60
Roosevelt, President Franklin D. 143
Rouse, Frank 236
Ryder, Orlando W. "Doc", Maj. Gen. 66

S

Safí 16
Salerno 103–106
Sbeitla 57–59
Schepers, Robert W., Capt., USN 236
Sicily
  Capo d'Orlando 95
  Catania Plain 93
  Gela 94
  Licata 87–91, 94, 99
  Messina 93, 96
  Termini Imerese 103
Skopje Earthquake 235
Smart, Jacob E., Gen. 236
Spaatz, Carl, Gen. 25–26, 52–53, 223
Spokane 241–242
Stalin, Joseph 203
Stetson, Loring F. "Stets", Col. 120, 132, 135, 141
Sullivan, John E., Capt. 183, 186

T

Tedder, Sir Arthur, Air Marshal 52, 84
Thacker, Robert E. Col. 183–184
The Citadel 196
The Hump 136
Thelepte 38–39, 43–44, 49
Third Air Force 161
Tony Romano 99
Truscott, Lucian, Gen. 95
Tunis 74–76
Tuskegee Airmen. See 99th Fighter Squadron
Twelfth Air Force (U.S.) 60, 94, 100, 111

V

Vairano Scalo 112

Vance, Reginald C., Col. 209
Vichy French 23, 36–37
Villa Atina 120
Volturno 111–112
von Arnim, Hans-Jürgen, Gen. 43, 60, 73

W

Wallace, Vice President Henry A. 158–159
Wang, S.M., Gen CAF 215–216
Watts, Daniel, Lt. Cmdr. USN 167
Watts, Helen "Jeanne" 167, 190, 198–200, 242
Western Task Force 15–16
Wheeler, Earl G., Gen. 235
Wolfe, Thom 184
Wright Patterson AFB. See Patterson Field; See Wright Field

Y

YB-49 Flying Wing 190
Yeager, Charles E. "Chuck", Capt. 183–185
Youks-les-Bains 44

Z

Zagreb Floods 235